Simple PLUMBING

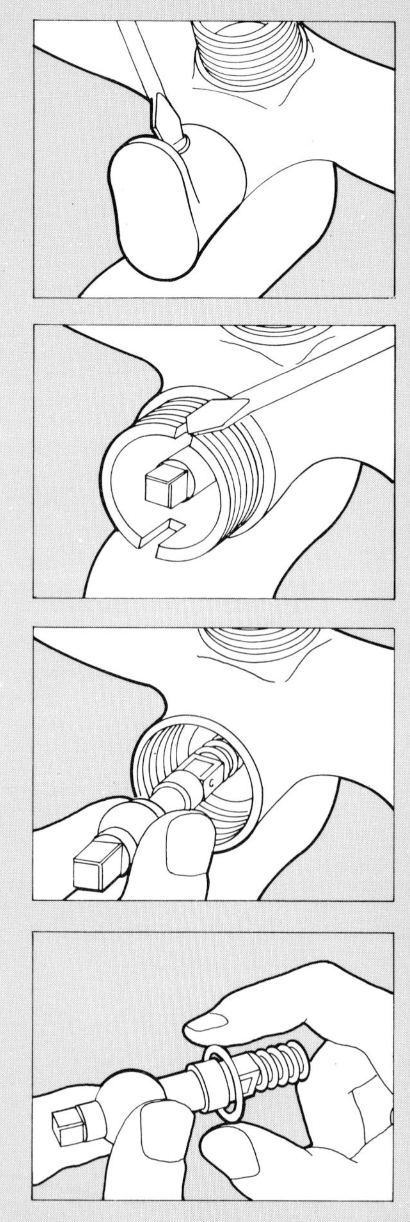

Simple PLUMBING

Elizabeth Gundrey

WARD LOCK
LIMITED LONDON

First published in Great Britain in 1978 by Ward Lock
Limited, 116 Baker Street, London W1M 2BB, a member of
the Pentos Group.

Designed by Mel Saunders

Illustrated by Mike Saunders

Cover photograph by Rob Matheson

House editor Gill Upton

Text filmset in Linotron Century 9/11pt by Morrison and
Gibb, Edinburgh

Printed and bound in Great Britain by Morrison and
Gibb, Edinburgh

British Library Cataloguing in Publication Data

Gundrey, Elizabeth
   Simple plumbing – (Household repair series)
   1. Plumbing-Amateurs' manuals
   I. Title
   696'.1   TH6124

   ISBN 0 7063 5431 1

Acknowledgements

The publishers would like to express their grateful
thanks to Angel Electrical and Plumbing Services,
83 Goswell Road, London EC1 for help in the preparation
of the cover of this book.

Contents

Contents

Contents

Introduction

For those of little skill

To call a plumber to one's home these days costs
several pounds, no matter how trifling the job that
has to be done. This is particularly hard on people
of limited means – and it is quite unnecessary to be
a 'helpless little woman' (of either sex!) in the case
of many repairs that involve neither technical
knowhow, nor much muscle.

This book is about repairs that can be
undertaken by a complete amateur. They call for
no previous experience or skill, and need a
minimum of very ordinary tools. The book is
written in everyday language, avoiding technical
jargon wherever possible.

The main essential is to have the *right* tools –
few of them but well chosen (see pages 16–19). For
example, nuts and screws yield much more readily
to spanners and screwdrivers that are of the right
size.

To use the book, first define what has gone
wrong and look this up under the heading
'symptom'. There may be several possible causes:
only those an amateur can tackle are dealt with in
this book, and if none of these is responsible for
your particular problem you should call in a
plumber (see page 11).

Tenants and leaseholders

Unless you own the freehold of your house, some
repairs may be the landlord's responsibility rather
than yours. This is particularly likely to be the
case in flats where certain services are common to
more than one home. Examples are drains, the cold
water supply, gutters and downpipes. It is the
landlord who should keep these repaired. If he fails
to do so, any Citizens Advice Bureau or Legal
Advice Centre will tell you what to do. Be careful
about undertaking any repairs to such things
yourself: in particular, you would be liable for any
damage you might do.

Introduction

Fixtures within your home are also the landlord's property but an obligation to keep them in good order is normally the tenant's: examples are baths, basins, sinks, WCs, taps, and water heaters. If you go further and put in improvements, these become the landlord's property: examples are new taps or WC seats – unless you were to replace the old ones when you move out.

Your insurance policy

You (or your landlord) should have a policy covering the building, and you should also have one that covers its contents. One policy may combine both. Your building policy will be in the hands of your building society if your house is mortgaged.

Such policies cover fire, flood, theft and some other disasters too. You may be able to claim for the cost of putting right the following plumbing mishaps:

Flooding from cracked tank or pipes. (Damage to carpets, or even dry rot, may qualify for a claim, but not replacement of the tank or pipes. Keep the damaged goods as the insurance company may want to inspect them.)

Flooding from cracked basin, WC etc. (The replacement of these will probably be covered, but not if you caused the crack while carrying out repairs.)

Flooding from blocked drain, rusted gutters etc. (Less likely to qualify because blockages are usually due to negligence).

Flooding from a neighbour's property. (Leave it to the insurance company to claim against him. Your policy should cover you against any claim if your water enters a neighbour's premises.)

Damage to underground pipes is covered in some cases.

Damage to landlord's fixtures (a contents policy may protect tenants against claims or the cost of repairs).

Introduction

Calling in the professionals

Unless a plumber belongs to a firm, it is best to phone outside working hours. Get an estimate for any small job before he starts: this is not so binding as a written quotation, but is a good deterrent to overcharging later. For big jobs, get written estimates from several – one might quote twice as much as another, but might be providing a more elaborate or better quality job, so read and compare descriptions carefully. Those who advertize emergency service are apt to be expensive. If you have a complaint that cannot be resolved, take it to the Consumer Protection Department of your local Council or to one of the trade associations named below if a member of theirs is involved.

Plumbers In an emergency, the water authority or even the police may give you names of plumbers. Some water authorities will not only turn off the supply but even change tap washers or ball-valves (free or at a moderate charge).

Fully trained plumbers can be located by asking:
Institute of plumbing, Scottish Mutual House, North Street, Hornchurch, Essex.
(Hornchurch 51236)
or National Association of Plumbing Contractors, 6 Gate Street, WC2A 3HX.
(01-405 2678)

Hiring gear – big or small

To find a firm that will hire you anything from a screwdriver upwards, look in the Yellow Pages of the telephone directory under H, C or P (for 'hire', 'contractors' or 'plant') or in the National Plant Hire Guide which many libraries have. The Hire Association, 12 Voluntary Place, Wansted, Essex, can supply addresses. Terms vary enormously. Not all firms give instructions for using complex equipment, and it's wise to check the condition it's in and the length of flex if it's electric. The Hire Association, 12 Voluntary Place, Wansted, Essex, can supply addresses.

The cold water supply

Whose responsibility?

The Water Authority is liable only for the public supply running beneath the road and its responsibility ends at the stopcock through which the water enters your service pipe under the ground. This stopcock, usually in a small pit below the pavement but possibly in your garden or cellar, needs a special Water Authority key to turn it off and on. The Water Authority is also responsible for sewers, although the actual work involved in looking after them may be done by the local council.

The main service pipe

Normally this gives no trouble unless gardening operations have left it with much less than a metre of soil above, in which case frost may crack it. It ordinarily enters the house via the kitchen floor (at this stage it is usually referred to as the rising main). If it is not embedded in a solid floor but exposed to the cold air beneath floorboards, it should have been insulated against frost when it was installed.

Turning off the main supply

You should be able to see the rising main under the kitchen sink or elsewhere and, just above floor level, there is usually a stopcock on it which can be turned by hand if the entire water supply needs to be cut off: for instance, if a pipe or tank leaks; while carrying out certain repairs; or when leaving the house vacant for a long period (insurance policies can be invalidated by failing to do this). Above the stopcock may be a drain-cock, operated by a spanner, the purpose of which is to empty the rising main, an operation rarely needed. It has a ridged nozzle so that a hosepipe can be attached.

If you drain off the water supply, also turn off or rake out the boiler or water heaters before these run dry. When turning the water on again, be sure

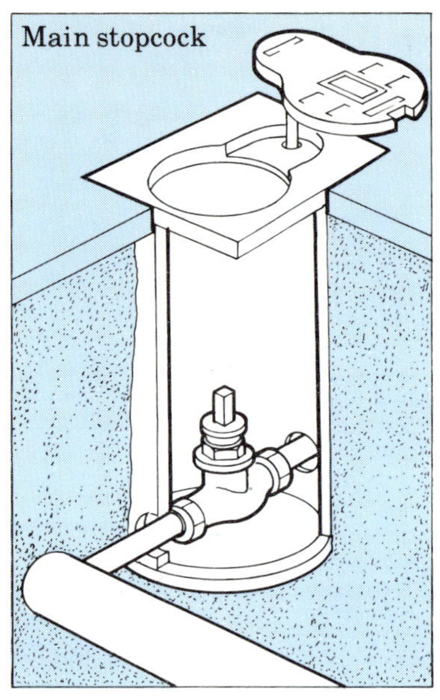

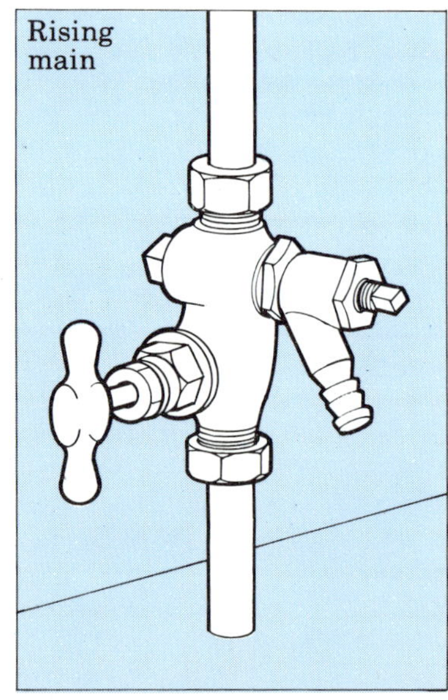

The cold water supply

to turn really full on – otherwise water pressure will be reduced, with an adverse effect on any shower, automatic washing-machine, the flushing of the WC, etc. Because stopcocks can become immovable through prolonged disuse, it is a good idea to turn them off and on again occasionally and to grease their stem. (For methods of releasing stiff ones, see page 28.)

To empty the water system it is, of course, necessary not only to cut off the incoming supply but also to turn taps on and flush the WC, in order to drain the indoor pipes.

Indirect supply

Although in some homes all cold taps and WCs are supplied straight from the rising main, today most Water Authorities allow only a kitchen cold tap and possibly the non-storage type of water heater (such as Ascot) to be supplied by branch pipes from it, with all the rest of the water carried up to be stored in a cold water tank (usually in the attic and occasionally on a flat roof, because the higher it is the better will be the water-pressure and the less likely is its noise to be heard in the house). The reason for this indirect method is largely because it cuts down any risk of contamination getting drawn back into the public supply. It also keeps water pressure constant in the house, and ensures a supply for a while even if the mains water is cut off.

From this tank will descend at least one pipe, with branches serving bathroom taps and WC, while a second will serve any storage-type water heater (such as an electrically heated cylinder). A third may serve a shower. These pipes may have their own isolating stopcocks, usually near the tank, so that if you want to drain only, say, the cold supply to the bathroom without affecting the hot supply, you can do so. A large screwdriver may be needed to turn these.

If these pipes do not have their own stopcocks,

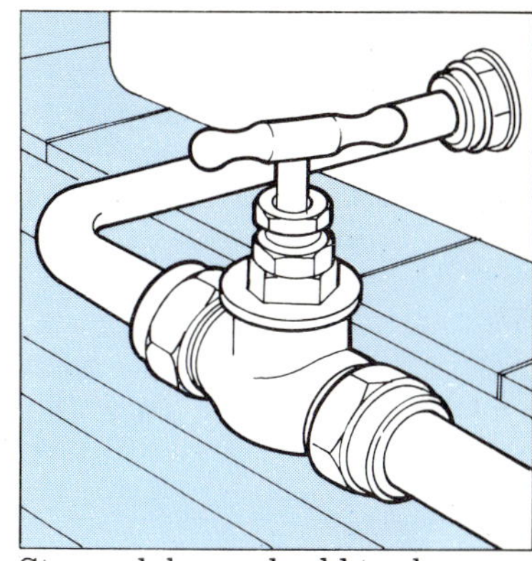
Stopcock beyond cold tank

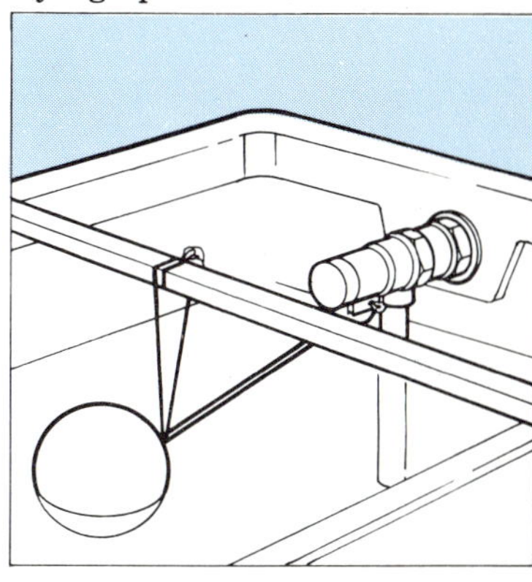
Tying up ball float

The cold water supply

you can cut off the supply to the cold tank by either using the main stopcock under the sink or tying up the arm of the floating ball in the tank, which will close its intake pipe. Alternatively (to save having to empty the tank too) use a broomstick and cloth to plug the outlet from the tank.

Emergencies: Put the telephone number of your Water Authority here:

....

and your plumber: ....

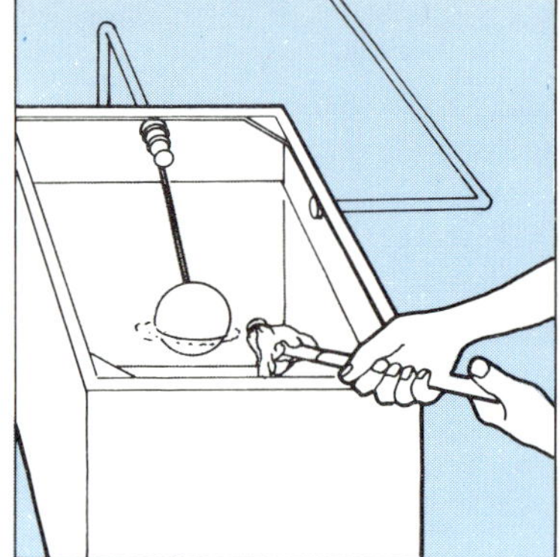

Plugging tank outlet

Access to the cold tank

Because so many plumbing repairs involve turning off the water-supply from the cold tank in the attic, it is necessary to have a convenient ladder and, if necessary, to put boards across some of the joists so that you do not put a foot through the ceiling below.

There are special loft ladders which are kept folded in the attic (screwed or bolted), to pull down when necessary. If well made, these are safer than loose ladders and naturally cost more. Sliding aluminium ones tend to be better than the concertina types or wood ones.

Points to check when buying one are: is it long enough; how much landing, loft and hatch space does it need; how easy is it to get it down – here spring-assisted ones score; has it got a hand-rail; has it wide and flat treads or merely rounded rungs? (See page 82 about ladders.)

Keep sliding sections lubricated.

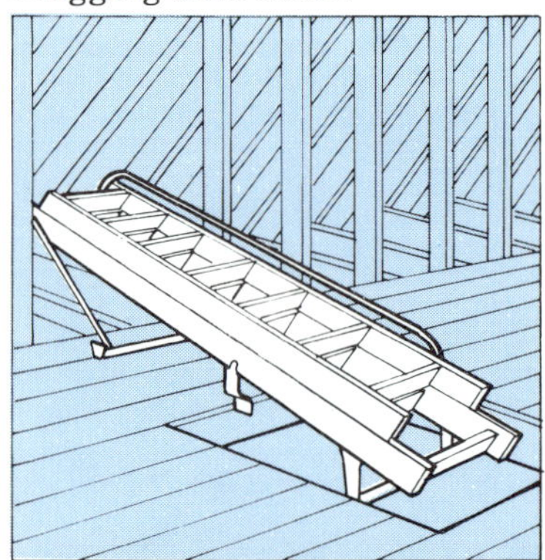

Attic ladder (stored)

Attic ladder (pulled down)

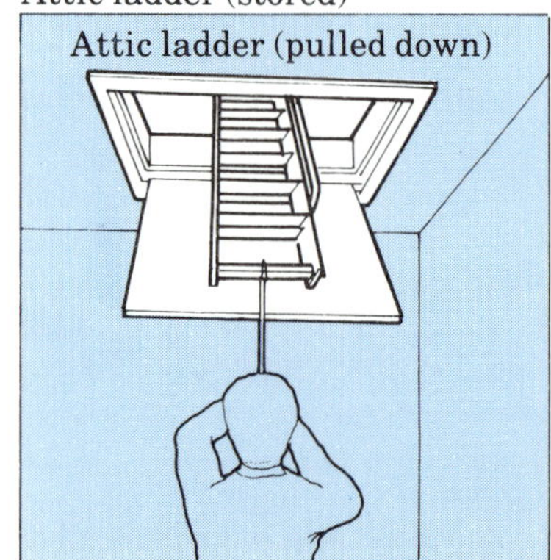

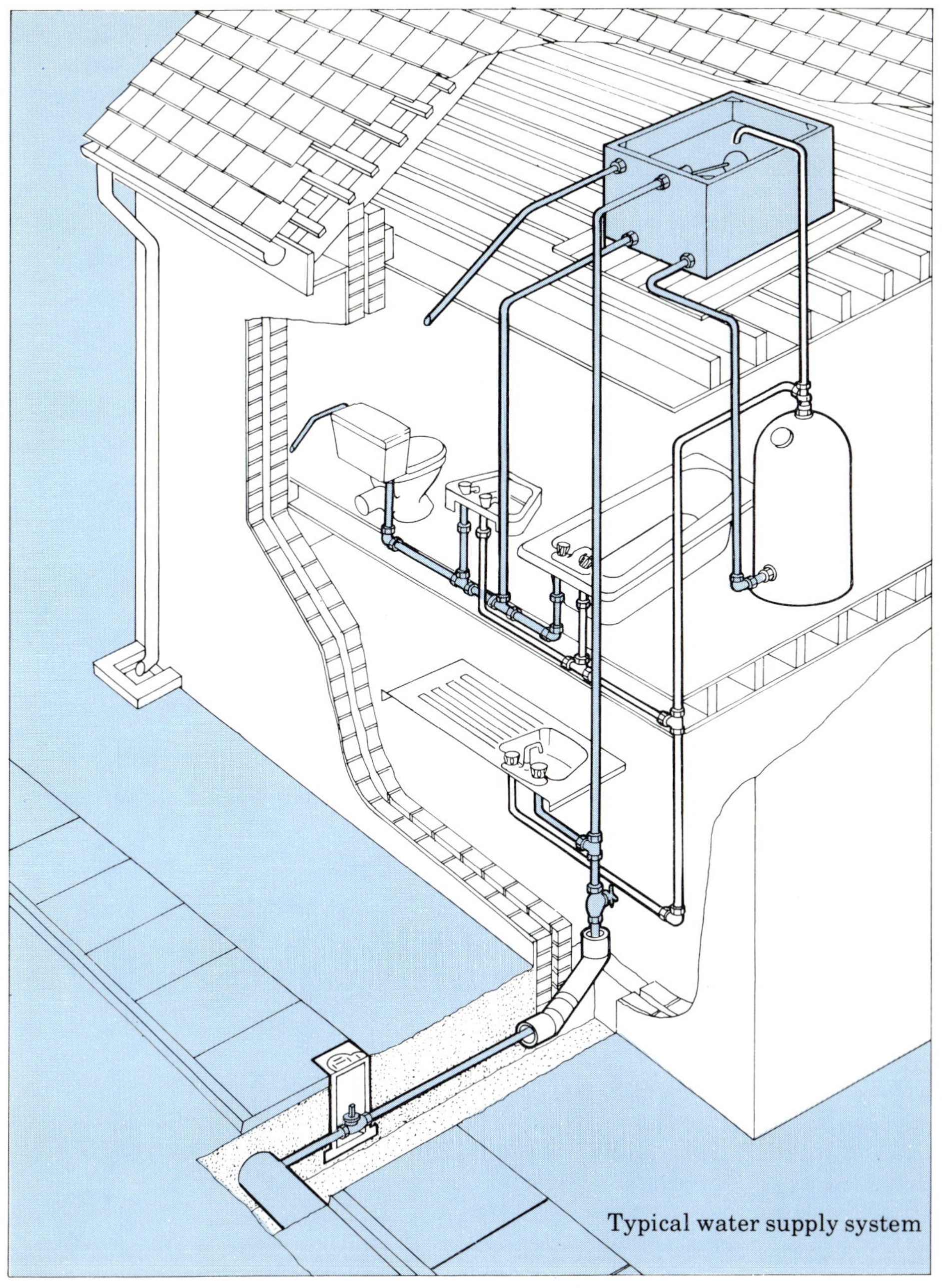

Typical water supply system

Tools and materials

Undoing connections

The hardest part of plumbing repairs is often getting things undone. With prolonged disuse, screw-threads are apt to seize up, particularly if rust, hard-water scale or paint have worsened matters. Tapping, or pouring boiling water over them, may help. Paint stripper can be used, or a blow-lamp if adjacent surfaces are protected with wet cloths.

Choosing the right tools is essential. Spanners must be exactly the right size for the job, and must grip without slipping. The wrong tools can spoil the shape of nuts, maul and distort pipes, and damage chromium. These pages show the choice available, and which to use for which job. (To undo anything, turn anti-clockwise: right-handed people should pull towards themselves, left-handed people should push away from themselves.)

To help with really stubborn cases, there are dismantling chemicals ('Plus Gas Formula A' or Rocket WD40) in an aerosol. After applying one of these, give a tightening-up movement to crack any rust or scale before attempting to undo the part. If this does not suffice, give another application and allow it to soak in for a few minutes.

If all else fails, a flexible tapered saw blade may cut through a bolt. To save the bolt while cutting off the nut, it may be worth buying a nut-splitter; turning the screw at the top with a spanner exerts such force on a cutter that it will split the nut in two.

Prevention of stiffness

When reassembling things like taps and waste-traps, grease the screwthreads with Vaseline to keep them moveable in future.

Where water pipes are concerned, the traditional method is to apply a special compound ('Boss White') and hemp to the screwthread before tightening it. This seal remains sticky and does not

Tools and materials

harden, so that the joint can be unscrewed at a future date if need arises. Plumber's Mait is another non-setting sealing compound for this purpose. A new alternative involving less mess is to wind round the screwthread a fine tape made with PTFE (as used for the non-stick coating on saucepans). This makes an absolute seal but easily comes away if the joint has to be unscrewed later (apply the tape in the same direction as the screwthread). Plastic joints, though needing no sealing, have screwthreads which tend to jam if not turned carefully: do not force them.

Spanners

Hexagonal nuts need spanners to fasten and unfasten them. Because nuts come in a variety of sizes, some spanners have adjustable jaws. They are sometimes referred to as wrenches.

Fixed jaw spanners may be:
Open-ended: usable only if you can approach the nut from the side and have plenty of turning space.
Ring: this kind, placed over the top of the nut, is very effective. It needs to be given twelve little turns to revolve the nut completely (therefore less turning space is needed).
Sometimes spanners have a ring at one end and an open-end at the other.
Ring heads also feature in socket spanners, where they are put on a tube with a choice of handles at the other end. A socket spanner exerts more energy, and with it even awkwardly placed nuts can be reached.
Sizes Unfortunately, several systems of nut sizes are in use in Britain, and often neither they nor spanners are clearly marked. The only way out of this dilemma is to take the nut or its measurements to the tool shop (or motor accessory shop) when you go to buy a spanner. Do not buy spanners in sets containing many sizes you will never need. You may find it helpful to have

Tools and materials

duplicates – one to hold and the other to turn.

Quality Over-long (and over-thick) spanners can be a nuisance in confined spaces, although the longer the spanner the more energy you can exert on it. Open-ended spanners are easier to use if their heads are at a 15° angle.

Adjustable spanners (wrenches) Although there are many patent versions, the simple crescent-jawed type remains the best for most purposes. A thumb-screw moves one of the jaws until the right size is secured – up to 20–30mm, depending which model you buy. Check the ease of dismantling and cleaning, because rust and dirt can clog the adjustment (dismantling solvent will help); and avoid over-thick spanners that will be difficult to use in confined spaces.

If you need to deal with larger nuts and pipes, a square-jawed spanner may be necessary but this needs more space in which to operate. Stillsons are a version of this in which the jaws are serrated and the upper one is spring-loaded.

Pipe wrenches are not suitable for use on nuts.

Finally, to get under basins and baths to undo taps there are specially angled wrenches.

Pliers

Pliers have square jaws to grip most things, a curved portion to hold pipes or nuts, and a cutting portion with which to sever wires: see diagram. (Some have another wire cutter too – a small groove on the outside.)

When choosing pliers, check that the cutting parts meet completely and that the pliers are fairly hard to close – they will slacken with use. Although 7-in (177mm) pliers are a popular size, smaller ones will be more convenient in confined spaces. Thin jaws are often handier, too. Where plumbing jobs are concerned, check that the curved part will accommodate pipes of up to, say, 20mm. Feel whether the size and shape suit your hands. Ribbed handles are less likely to slip than

Tools and materials

smooth ones, and check also that the serrations on the jaw go right up to the tip: important when pulling out split-pins, for instance.

The Mole is a cross between pliers and an adjustable spanner. Once in position, its jaws may be locked so that you can relax your grip. One version has curved jaws for holding pipes.

Screwdrivers

Types of tip Ordinary screws call for a plain-tipped screwdriver, but you may come across some recessed screwheads that need a screwdriver with a cross-shaped tip. (There is a screwdriver that has a reversible tip – one end of it plain, the other with a cross.) Unfortunately, there are two kinds of cross-headed screws, Philips and Pozidriv, and each needs a different screwdriver tip.

Sizes Where plain screwdrivers are concerned, the width of the tip should correspond with the slot in the screw. There are some screwdrivers on sale which come with a set of interchangeable tips.

Design Although large round handles usually give the best grip, slender fluted ones are better when working with wet or greasy hands or in a confined space. When choosing a screwdriver, check that its tip is really square, its blade thick and its handle without sharp edges to dig into your hand. It is worth seeking out a shape designed to enable a spanner to grip it when extra force is needed. (Footprint make one.)

If you are prepared to pay more, you can get screwdrivers that will themselves hold the screw securely while it is being positioned; but you can improvise a screw-holder yourself by slipping a bit of rubber tube, drinking straw or even Plasticine, over the tip of the screwdriver.

There are offset (angled) screwdrivers designed to get into awkward corners.

Easier to operate are ratchet screwdrivers and, being usable with one hand only, they are helpful when working in confined spaces.

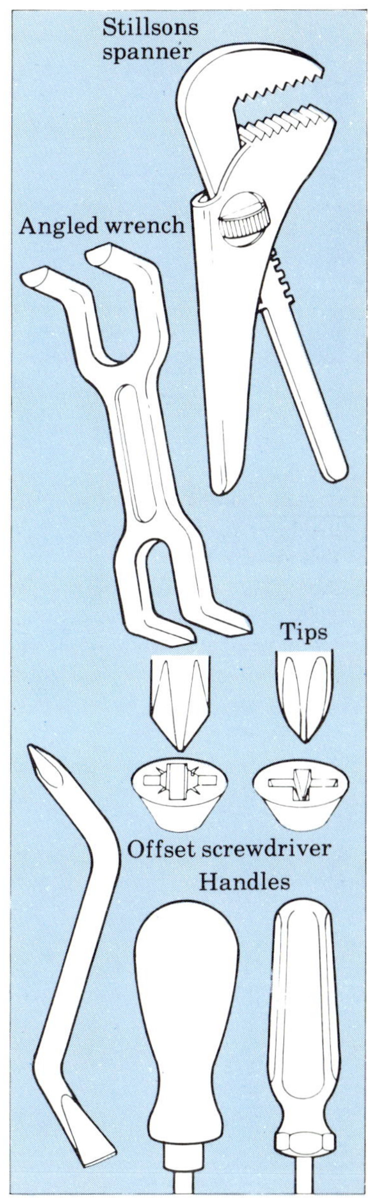

Tools and materials

Most ironmongers will regrind worn screwdriver tips.
Removing a screw (especially if the slot is damaged)

Heat the head of the screw with a blow-lamp. (Screw heads must be clean.) Or use a soldering iron. Put another screw on it upside down. When cold, this can be gripped by pliers, to turn it.

Fillers for gaps and cracks

There are now so many fillers that choice is difficult. For a start, be clear what characteristics you need from a filler.

The repair of leaks in pipes, gutters, tanks and so on obviously calls for waterproof materials, which must also be heat-resistant if hot water is involved.

In the case of WCs or wash basins, any filler must be impervious to germs – nothing at all porous or absorbent will do.

In some gaps, such as those between a bath and the walls, the filler must remain a little resilient: anything hard may crack because of the slight movement that occurs.

Bearing these needs in mind, the following are some of the fillers likely to be most useful for plumbing repairs – all are waterproof.
Epoxy putty (eg Sylmasta)
Strong, hard, smooth, heatproof. White (can be painted). Usable for cracks in pipes, basins, etc. See page 32.
Plastic steel (eg Devcon)
Epoxy plus steel particles. Very strong, hard, smooth, heatproof. Fine for metal pipes.
Leak sealer (Fernox LS-X)
Will adhere even to wet surfaces, seals cracks in any material, heat-resistant. Does not harden, so can be used inside joints that may later need to be undone. Transparent.

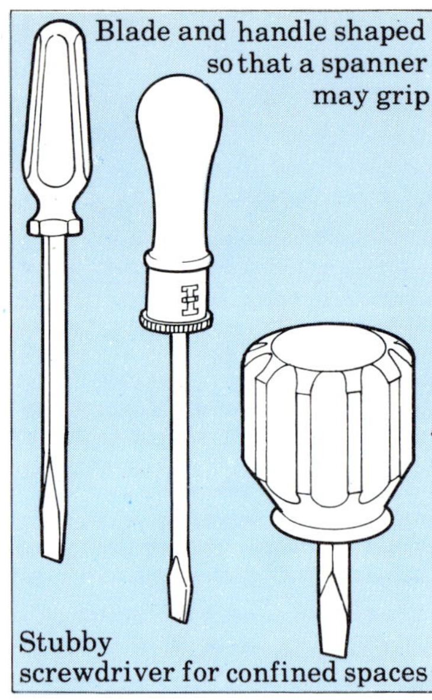

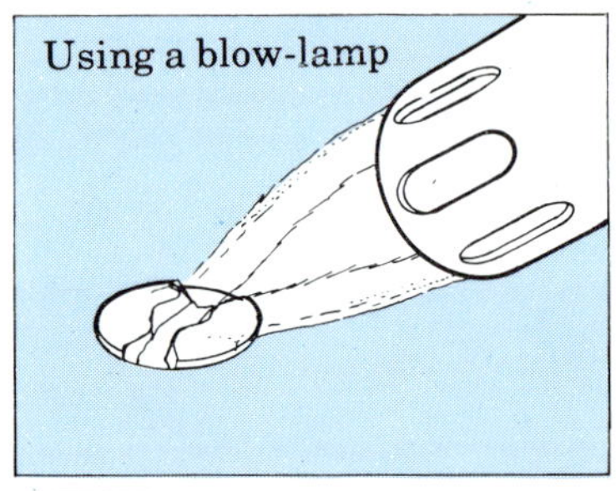

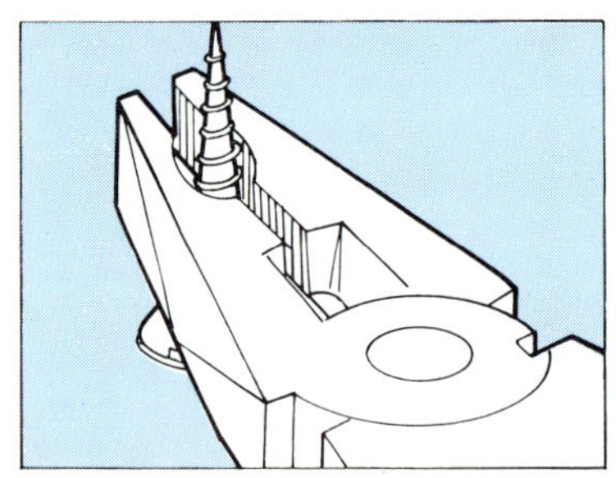

Tools and materials

Silicone rubber (eg Evo-Stik Colourseal)
Resilient, strong, smooth. White or coloured;
shiny. Best choice for gaps around baths and sinks.
Can even be applied under water.
Cement-based products (eg Moltofill paste, Rok-
Rap bandage – cement-impregnated)
Hard. Mainly for outdoor use. Not suitable for
plastic surfaces.
Polyurethane foam (eg Fomofill)
Sticky substance ejected from an aerosol slowly
foams up to fill even large or irregular gaps.
Resilient. Will even secure wobbling basins or
vibrating pipes, and fill spaces in walls where
pipes enter. Can be painted. Not suitable for
plastic surfaces.
Vinyl adhesive (eg PVX)
Specifically for cracks in things like plastic gutters
and rainwater pipes.
Non-setting mastics or sealants (numerous brands)
are usable for joints which may one day need to be
opened up again, but mainly for filling gaps
subject to some movement. Plastic-based ones are
white or grey; bitumen-based ones, for outdoor use,
are usually black.

Often waterproof building tapes are a good
solution to cracks and gaps. Some have a top
surface of aluminium foil. (More about these on
pages 32 and 74, Water-pipes and Gutters.)

Read labels before starting work, in order to check
three more points:
1 Is the filler suited to the surface (metal, plastic
   etc)?
2 What preparation is needed? (cleaning off dirt,
   grease, old paint or adhesive, moisture).
3 Any hazards? (Some fillers are flammable, some
   have noxious fumes.)

Note: Keep all chemicals out of children's reach.

Tools and materials

Rust removers and preventatives
To end this problem, many pipes, gutters and tanks are now made of plastic. But if replacement is not planned, iron or steel fittings will need attention.

Rust removal involves wire-brushing (with goggles on) or sandpapering followed by treatment with one of the chemicals that dissolve or neutralize remaining rust particles.

To prevent future rusting, it is essential to keep the surface coated with a rust-inhibiting primer underneath several layers of paint (after removing any rust already present). Alternatively, a rust-binding primer can be used which actually converts rust into a non-corrodible layer of iron phosphate.

Anti-rust chemicals may be in liquid, aerosol or jelly form, non-drip jelly being most useful on vertical or overhead surfaces. Bitumen paint (usually black) itself gives protection against rust but be sure to get an odourless grade if for use in a tank containing drinking-water. Galvanizing paint is an alternative: it deposits a new coat of zinc over the surface.

Keep flammable, acidic or noxious chemicals out of children's reach.

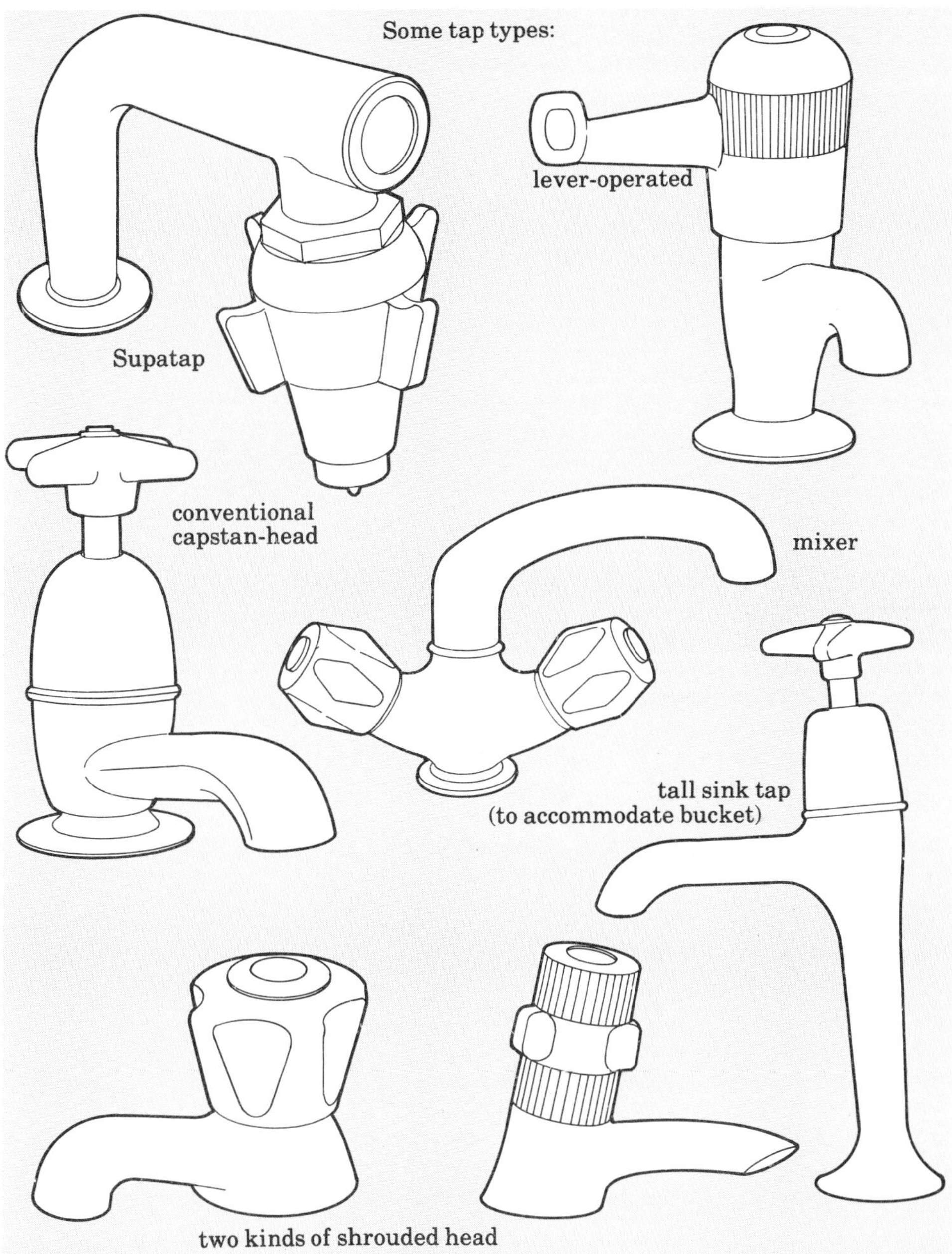

Some tap types:
Supatap
lever-operated
conventional
capstan-head
mixer
tall sink tap
(to accommodate bucket)
two kinds of shrouded head

Taps and stopcocks — dripping

Symptom
Continual dripping (Note: many Water Authorities will repair dripping taps without charge)
(Supataps: see page 27)

Cause
Washer deteriorated

Tools and materials needed
Washer: synthetic for hot or cold; rubber for cold only $\frac{3}{4}$in or 20mm for most bath and mixer taps, $\frac{1}{2}$in or 15mm for most sink and basin taps
Small spanners
Probably large spanner and dismantling lubricant (see page 17)

Method
1. Turn off water supply as follows: for a kitchen cold tap, use main stopcock; for any other, see page 13. You do not need to empty the hot water cylinder when re-washering a hot tap because the water comes only from the top of the hot cylinder. Turn taps on to drain the pipes.
2. With the tap in the fully-on position, remove handle by loosening the screw at the side and then tapping the handle upwards. If it will not move, raise the tap cover and jam a piece of wood or large spanner below, then screw the handle down and lever it off with the wood. Hot water may help to get the tap cover unscrewed by hand.
3. If a large spanner has to be used to unscrew the tap cover, use sticky-tape or a cloth to protect the metal. Not all tap covers are screwed on clockwise.
4. Having removed the tap cover, use a small spanner to undo the large hexagonal nut (hold the body of the tap firmly). Remove the top gear.
5. Pull out the jumper (a brass stem): there may be a split-pin to remove first. Undo its nut. Replace the old washer with a new one of the same size.

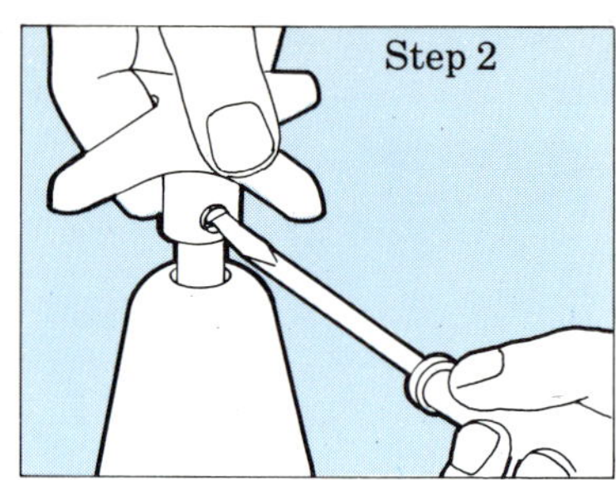

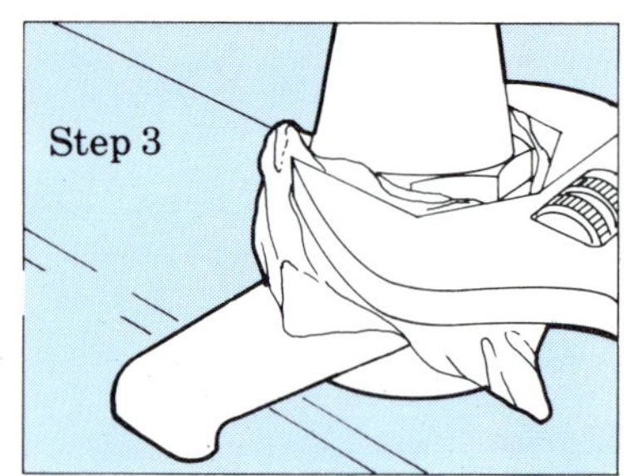

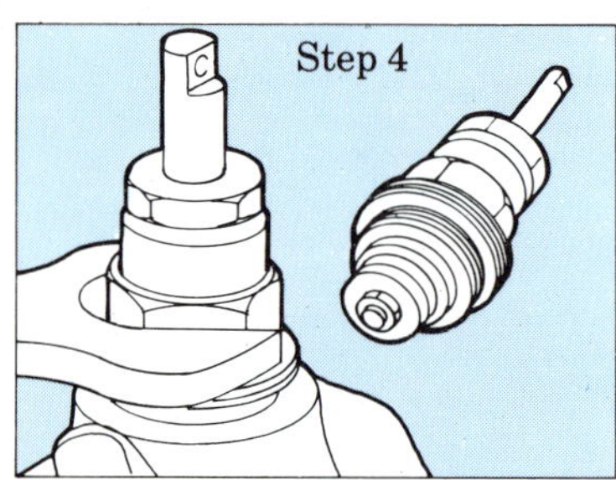

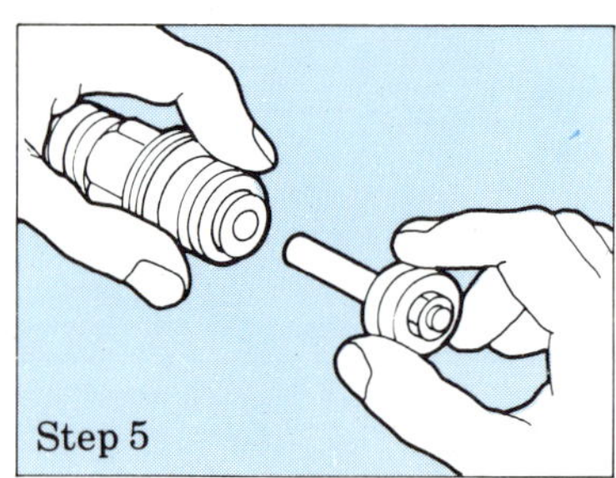

Taps and stopcocks — dripping

6 If even dismantling lubricant will not shift the nut, a new jumper complete with washer may have to be bought.

7 When reassembling the tap, grease all screw-threads (Vaseline or even lard will do) and do not over-tighten. Take particular care with plastic ones as it is easy to jam the screw-thread.

8 If dripping still continues, this means the valve seating below the jumper needs replacement. A plastic seating and jumper unit ('Full Stop') can be bought, to force down onto the existing seating. Screw the tap handle hard down.

Shrouded heads

To remove the combined cover and handle, prise out the plastic disc on the top to reveal a screw. When this has been unscrewed, a firm tug will pull the head off. Some have no screw but simply pull off; others have a screw at the side.

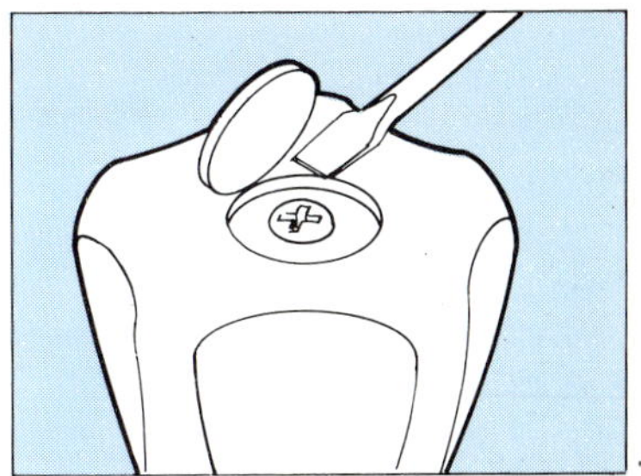

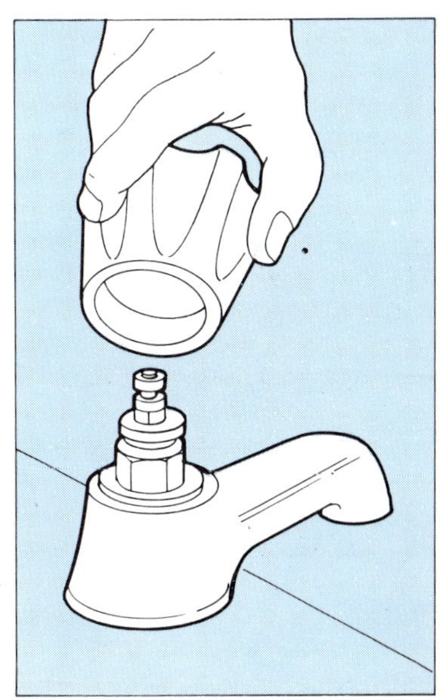

Taps and stopcocks — dripping

Symptom
Swivel nozzle dripping

Cause
Deteriorated washer

Tools and materials needed
Thin-nosed pliers
Washer(s)

Method
1 There is no need to turn off the water supply. Unscrew or lever up the shroud at the foot of the nozzle.
2 Use the pliers to take out the circular copper clip and slide it up the nozzle.
3 With the nozzle removed, replace the washer (or washers) with new ones. Wet the foot of the nozzle before replacing it.

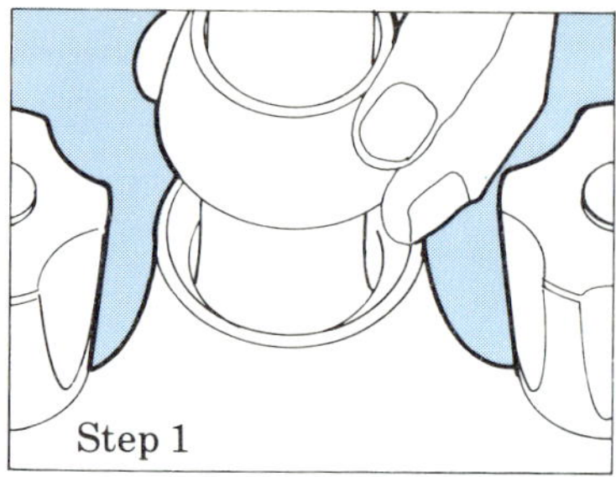

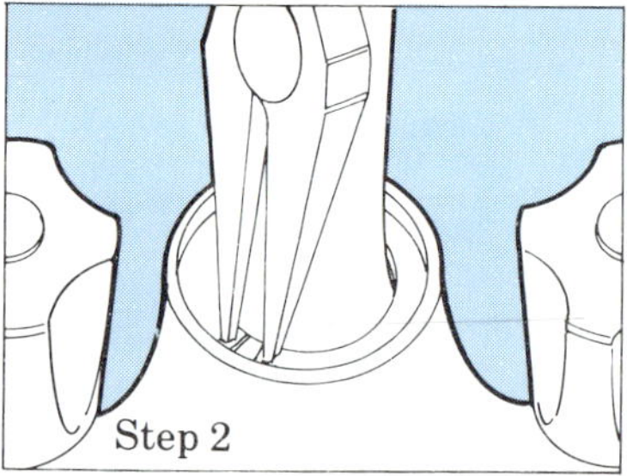

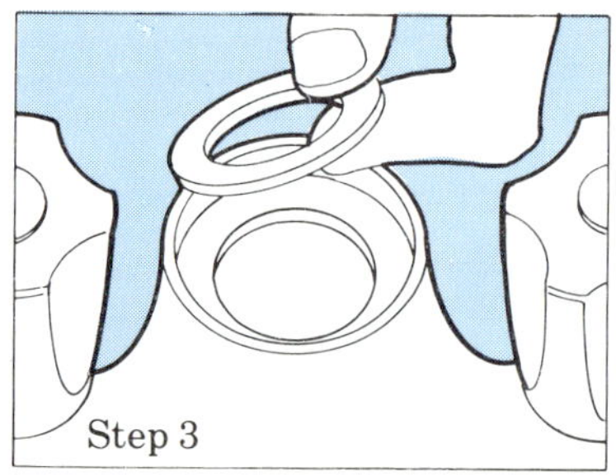

Taps and stopcocks — dripping

Symptom
Supatap dripping

Cause
Washer deteriorated

Tools and materials needed
Special Supatap spanner
Washer-jumper

Method
1 There is no need to turn off the water supply. Partly open the nozzle and unscrew the nut with the spanner. Detach the nozzle by unscrewing (the water will stop flowing).
2 Press nozzle on table-top to free the anti-splash device that is at the bottom of it, or push it out with a pencil.
3 Lever the washer-jumper out from the anti-splash device, brush the latter clean, and put in a new washer-jumper.
4 Put the anti-splash back in the nozzle and screw the nozzle back in place until almost closed (water will start to flow again). Tighten nut, then close the nozzle completely.

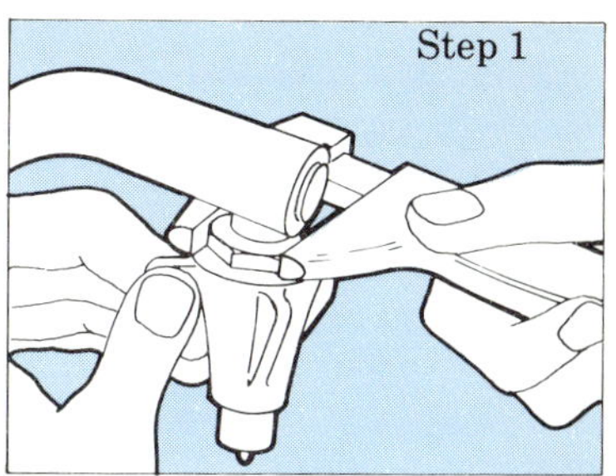

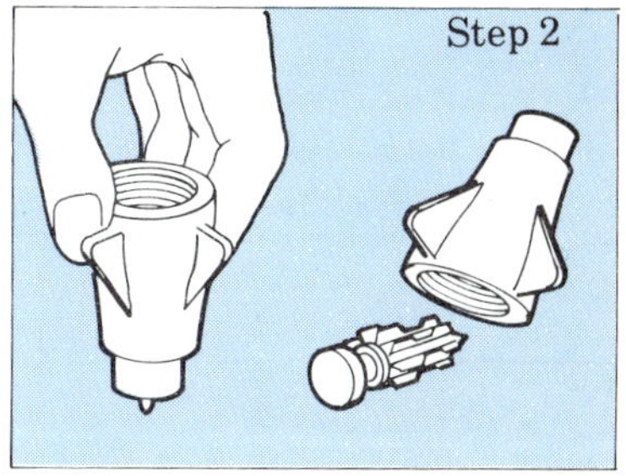

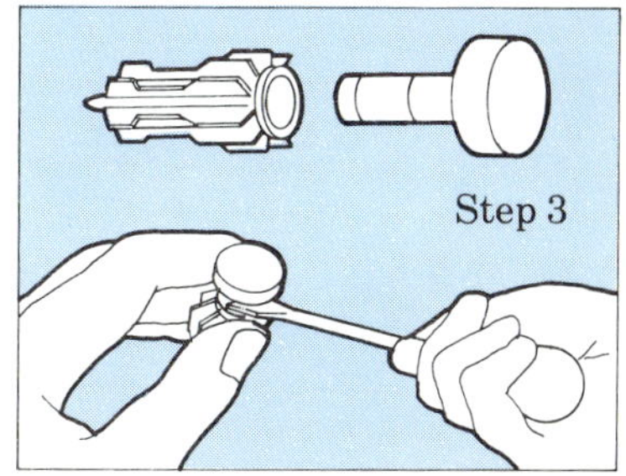

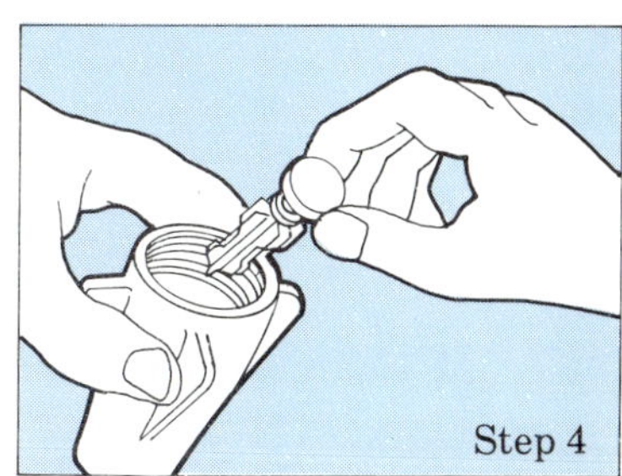

Taps and stopcocks — leaking

Symptoms
Leaking round the spindle
Tap hard to turn on and off
Tap spindle slips

Cause
Any of these may be due to deteriorated gland packing, or loosened gland-nut. (Gland packing may be affected by water forced back into the tap by a hose connected to it; or by detergent having entered from the top and washed the grease out of it.)

Tools and materials needed
Small screwdriver
Small and large spanners
New gland packing (i.e. string or wool, and Vaseline)

Method
1 With the tap fully on, remove the handle and tap cover (see page 24).
2 Turn the gland nut (at top) clockwise half a turn, using the small spanner. Test, and if this has not ended the leak, try another half-turn.
3 If this does not solve the problem, remove the nut completely, pick out the packing and replace it with new string smothered in Vaseline. Wind this round two or three times (do not over-stuff) and press it down with the screwdriver tip.
4 Re-assemble the tap.

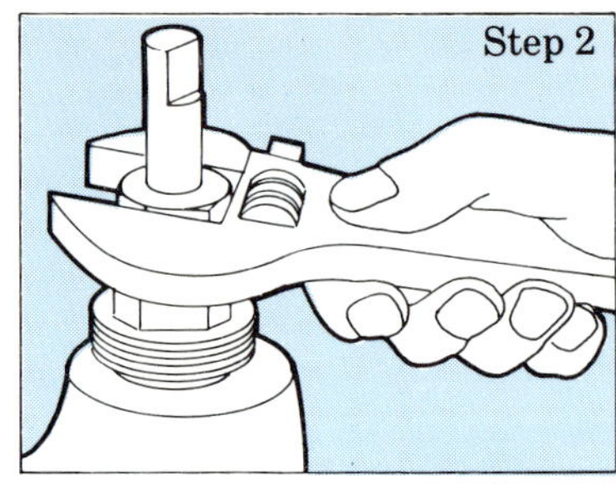

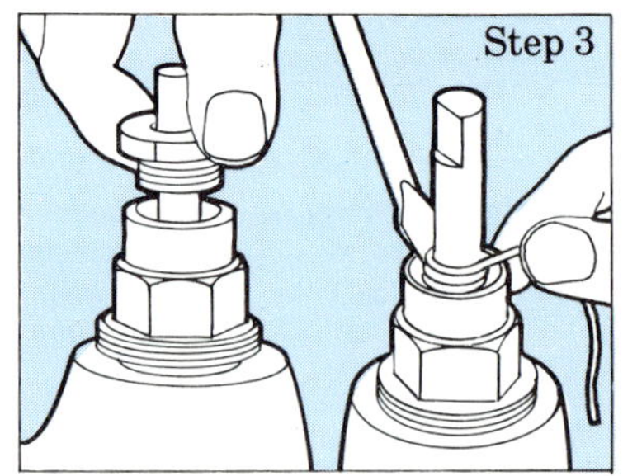

Some modern taps have a plastic O-ring on the spindle instead of gland packing, which is easier to replace than string.

Tap conversions

There are kits available for converting old-fashioned taps to shrouded heads, provided the taps are in good working order. They vary in price, style and ease of fixing. The one described here ('Aubit') replaces both the head and the spindle assembly too; some replace only the head. The latter type is a good choice if there is nothing wrong with the brass spindle. As only the head of the tap is removed, the water supply does not have to be turned off.

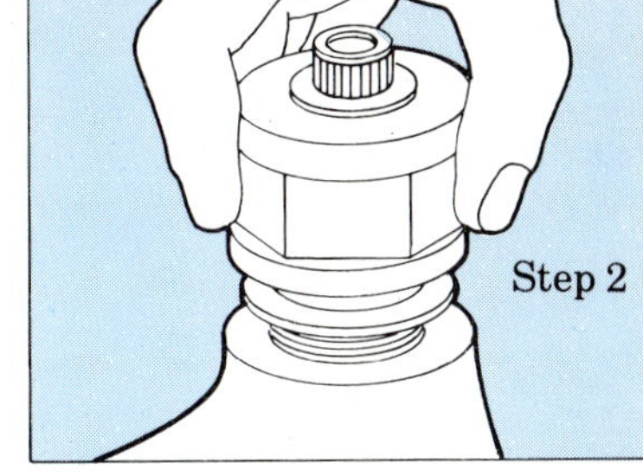

Tools and materials needed
Pair of tap heads
Spanners
Possibly dismantling lubricant

Method
1 Having turned water off (see page 12), unscrew tap cover and hexagon nut (see page 24).
2 Put new tap body in place of old one, securing it with a spanner – do not overtighten. (Keep the spindle retracted.)
3 Press head on.

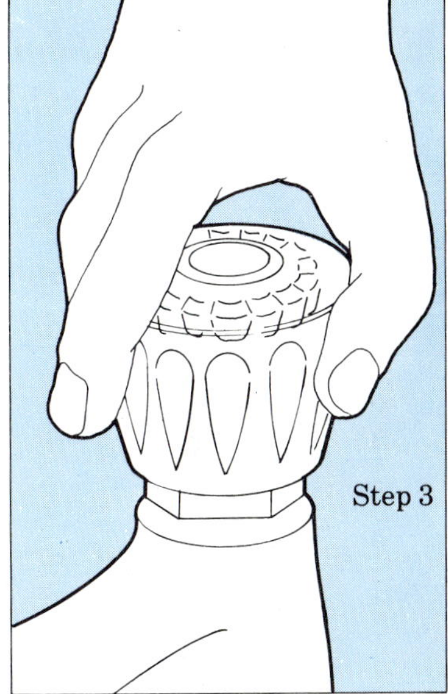

Tap replacement

Tools and materials needed
New tap
Washers (—see below)
Spanner
PTFE tape

Method

1 Having drained the water supply (see page 12) remove the old tap by undoing each of the two nuts below the basin. (If these are so awkwardly situated that an ordinary spanner will not reach, buy a special basin wrench as shown in the diagram.)

2 If the new tap is to go on a ceramic basin, place the washers above and below the ceramic before securing the tail of the tap with a nut.

3 If it is to go on enamel or steel sink or bath, it will need one ordinary and one 'top hat' washer, the latter to be fitted below the enamel rim as shown; secure tail with nut.

4 Before securing the lower nut which joins the pipe to the tail of the tap, wind PTFE tape clockwise round the screw-thread of the tail.

Note Opella make extra-long tap connectors, easier to instal. These plastic connections need no sealing round the thread.

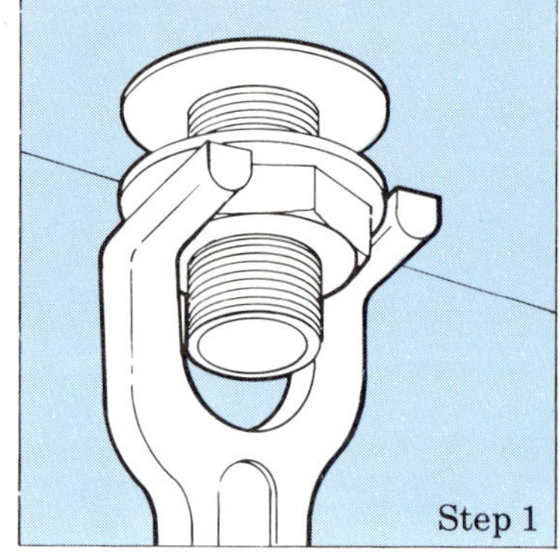

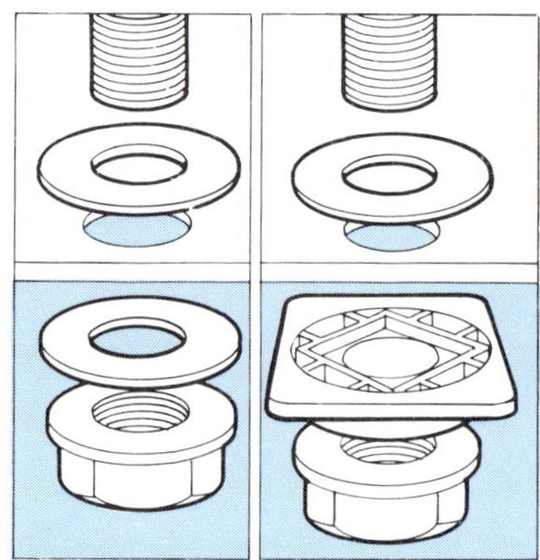

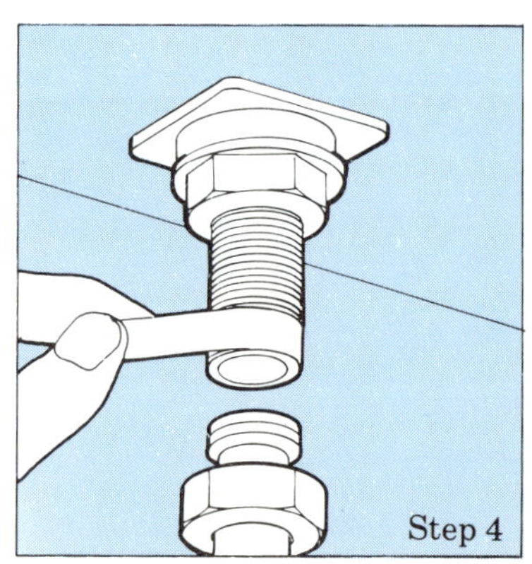

Stopcocks — leaking

Symptom

Stopcock dripping or leaking (this should be repaired without delay, for fear of dry rot)

Method

Repair in the same way as a tap – see page 24. In the case of a main stopcock, the Water Authority should be asked to cut off the water supply to the house – or to do the repair for you.

Water-pipes — leaking

Symptom
Leak

Cause
Crack, insecure joint

Tools and materials needed for cracks
Hammer (for lead pipe)
File or abrasive paper
Special tape or filler: see below

Methods for cracks
The following are a choice of temporary measures,
to use after draining the system (see page 12).
Usually a cracked pipe will need to be replaced in
due course (see page 79).

1 **First-aid** A small crack can be sealed for a short
period by rubbing soap into it and tying a rag
round; or by tying with a rag saturated in paint,
Vaseline or something greasy; or by securing with
plastic sticky-tape or a waterproof sticky bandage
(Band-Aid etc.). Many of the fillers and tapes
mentioned on page 20 will seal a crack. If the split
is in a lead pipe, it may be hammered together.
Restore water supply gently and at low pressure.

2 **Epoxy resin** (Isopon, Plastic Padding, Holts
Fibreglass etc.) This type of adhesive filler gives a
more durable repair. Rub the surface with a file or
abrasive paper, mix the resin in accordance with
the maker's instructions, and smear over and
round the crack. Bind a Fibreglass or other
bandage round and smear more resin over this.
Leave to set before restoring water supply.

3 **Epoxy putty** ('Sylmasta') is an even simpler
alternative. This comes in two parts which need to
be blended together by repeatedly rolling,
breaking up and kneading before it is applied. It
can then be smoothed down by wiping with a damp
cloth smeared with soap. A very shiny surface is

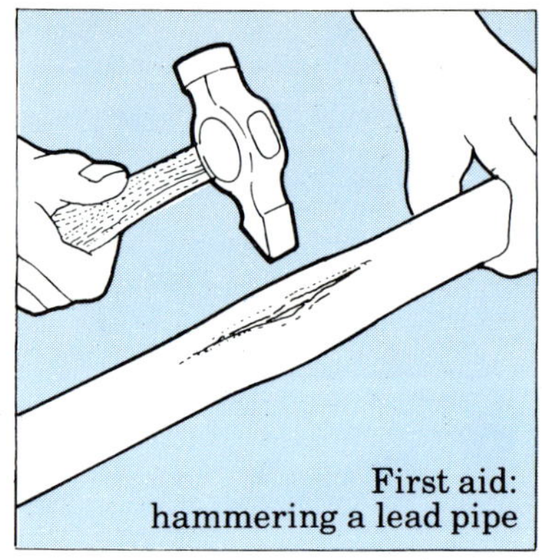

First aid:
hammering a lead pipe

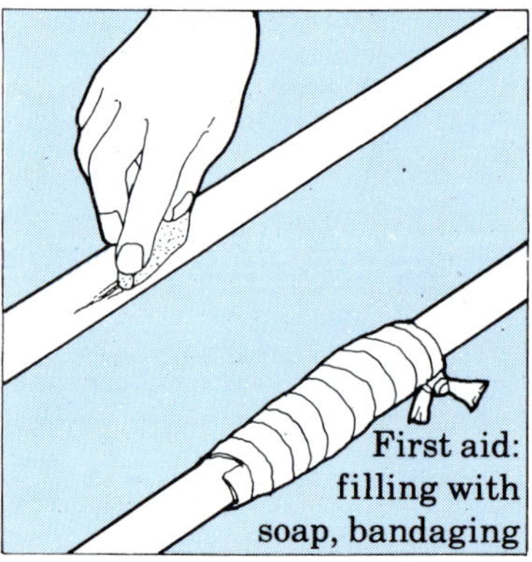

First aid:
filling with
soap, bandaging

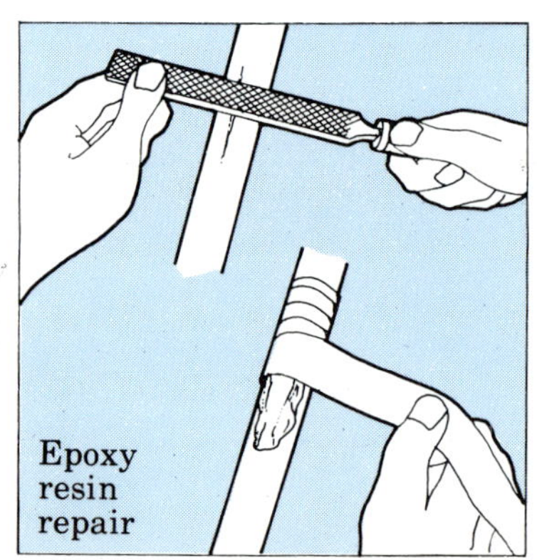

Epoxy
resin
repair

Water-pipes — leaking

left if the putty is covered with polythene (from a
plastic bag) while setting. It should be left for a
day before use to become as hard as stone, and can
be painted if desired.
4 Plastic steel (Devcon) is specifically for pipes:
epoxy plus steel particles.
5 **Pipe seal** (Rotunda) is a special 2-part tape, giving
a permanent seal. Wind the impregnated tape
round the pipe (or hose), then the reinforcing tape,
and then more of the impregnated tape.
Leaks
Two warnings:
1 Neglected leaks can start rot in woodwork.
2 If water leaks into electrical fittings, there is a
danger of shock: turn off electricity main switch.

Method for joints
See page 71.

Water-pipes — no water flowing

Symptom
No water flowing

Cause
Unless a leak, airlock or a defective tap is at fault, the cause may be frozen water. (Check whether, because water expands as it turns to ice, the pipe is also cracked or a joint has been pushed open, see page 32).

Tools and materials needed
Boiling kettle and cloths; or hair drier, fan heater, blow-lamp etc.; or hot water bottles; or candles
Salt (for waste-pipes)

Methods
1 Apply warmth to the frozen pipe wherever it is accessible; the heat will travel along the pipe to other parts. Start near tap, turned fully on, or near other outlet from which melting ice can escape. (If in doubt which part of the pipe is frozen, turn each tap on in turn and observe which ones run dry.)
2 In the case of waste-pipes and WC put salt in to thaw the ice, followed by boiling water. Watch out for possible leaks as the ice thaws.

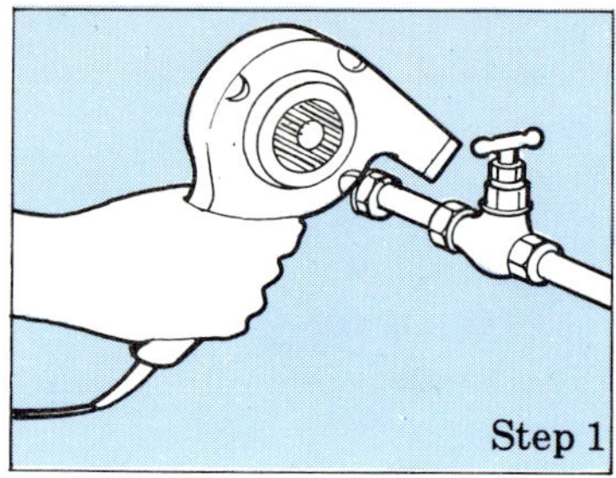

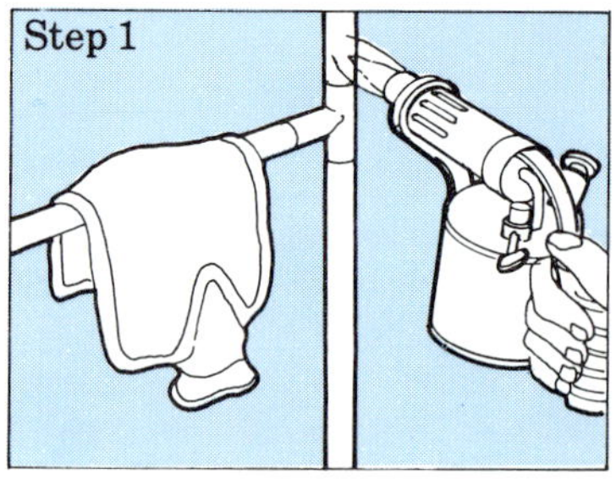

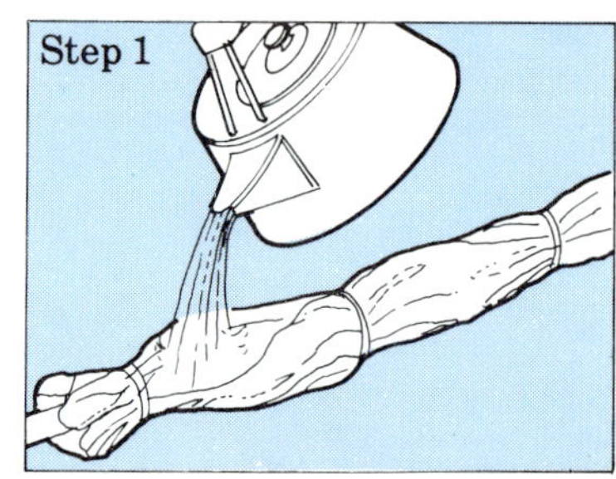

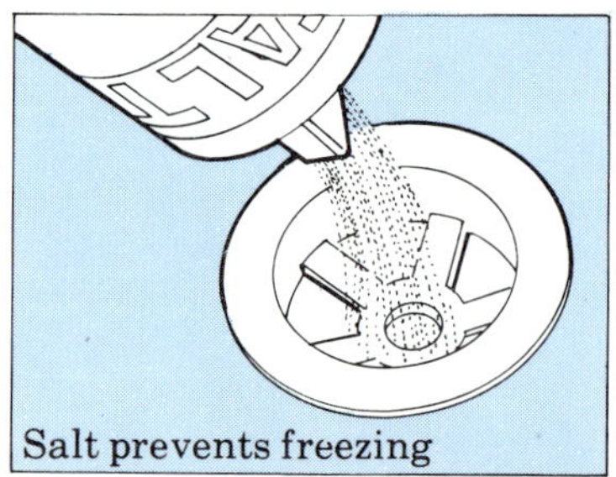

Salt prevents freezing

Water pipes — freezing prevention

Prevention of frost damage

Tools and materials needed
Lagging material
Scissors
String or sticky-tape
Salt (for waste-pipes)
Tank-lagging kit

Methods
1 **Pipes** Observe which are in vulnerable positions:
   on the outside or the inside of one of the outer
   walls of the house; or running through cold or
   draughty places (floor-space, attic or cellar). Bends
   and overflows are particularly at risk, and outdoor
   WCs.
   Either keep these places warmed during cold
   nights (even a light bulb hung close to a pipe will
   stop it from freezing) or drain the water system
   during cold nights or lag the pipes by one of the
   following methods (be sure also to lag any
   stopcocks, leaving only their handles protruding).
a Bandage-type lagging. Wrap round diagonally
   with edges overlapping, starting and finishing at
   right angles to the pipe. If there is a waterproof
   surface on the material, this should be kept on the
   outside. Secure ends with string.
b Flexible tubular lagging. This opens up to slip
   round pipes. It is vital to buy the size that will
   exactly fit the pipe. Cut with scissors into
   convenient lengths, slip on and then secure ends
   and any bends or joins with plastic sticky-tape.
   Occasionally check that the lagging is still dry:
   damp lagging is worse than useless. Lagging on
   pipes to outdoor taps is particularly vulnerable, of
   course, but can be protected with waterproof
   adhesive tape.
c Polyurethane foam (sprayed from an aerosol, see
   page 21). This is especially useful for inaccessible
   cavities.

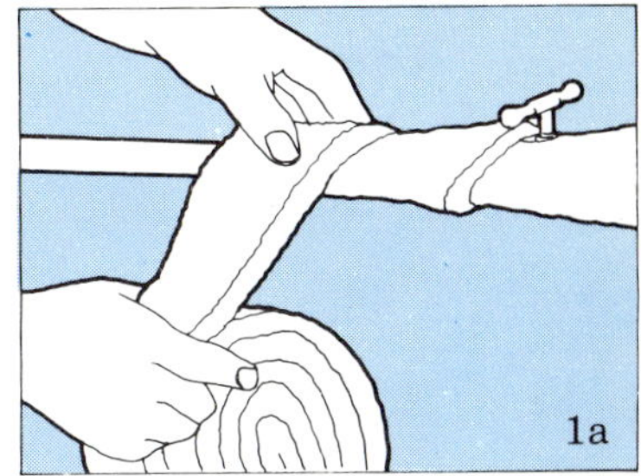

Bandage-type lagging

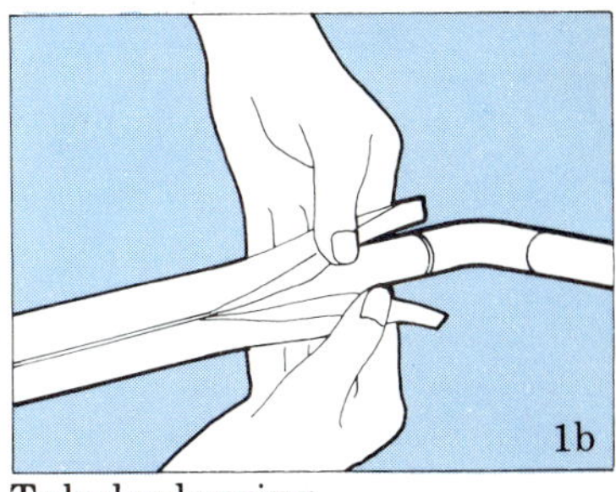

Tubular lagging

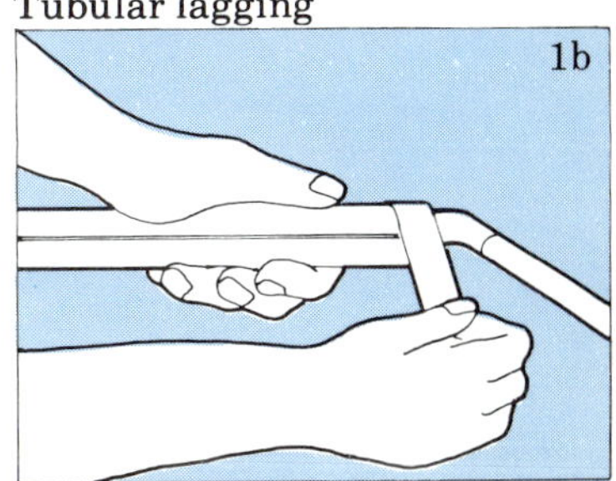

Water-pipes — freezing prevention

2 **Waste-pipes, WC and WC cistern** Add salt before nightfall, and put plugs in (unless you have a dripping tap).

3 **Tank** Lagging is important to ensure that water entering the pipes is no colder than necessary. Tank-lagging kits contain jackets, pads or boards of insulating material, plus instructions. Two good alternatives are:

a Fibreglass matting. If possible buy a roll as wide as the tank is deep, paper-backed for ease of handling. Wind round the tank, allowing an overlap. Cut slits with scissors for any pipes serving the tank. Tie on with string, but do not pull tight. Lay a loose piece on the lid. Do not insulate below the tank because this would cut off warmth rising from bedrooms.

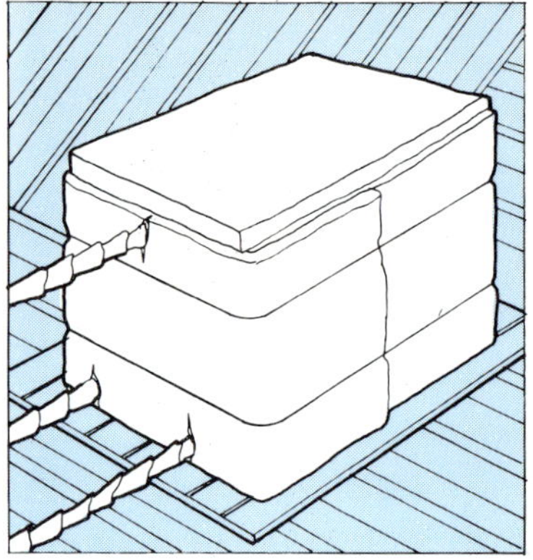

b Expanded polystyrene boards. Cut five rectangles, slightly larger than the tank sides and top, and with holes for any pipes. Join the sides with the kind of adhesive sold for polystyrene ceiling tiles, or with plastic sticky-tape; but leave the lid loose. On frosty nights, open trap-door so that some warm air can rise into the attic.

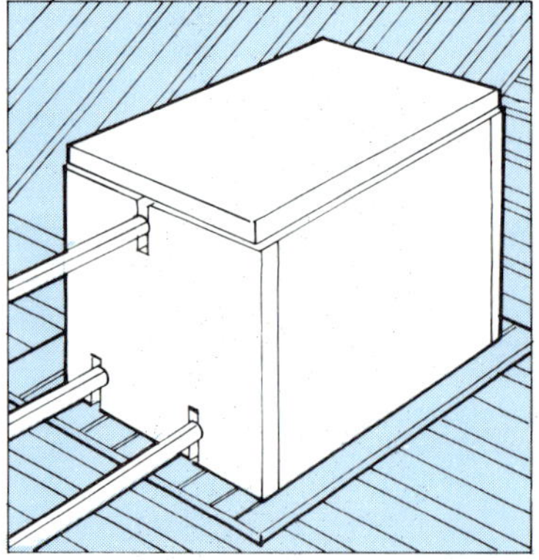

4 **Overflows** (from bath, WC cistern etc). These can be temporarily plugged with cottonwool to prevent icy air entering at night, provided you remember to remove the plug in due course. Special flaps for permanent attachment can be bought (made by Shires).

5 **Boiler** There is no need to let the fire out at night. Only if the house were left unheated might the boiler pipes freeze: in such an event, the fire must not be lighted nor the hot taps be turned on until the ice has thawed.

If a central heating system is left full of water in an empty house during winter, anti-freeze can be put in: Fernox is specially formulated for this.

Water-pipes — noisy — poor flow

Symptoms
Hammering or other noise
Poor flow

Cause
Faulty gland packing in a tap (see page 28); or faulty ball-float (see page 42). In the case of poor flow, a stop-cock may not be fully turned on. Some noises are due to pipes being inadequately secured to joists or walls.

Water-pipes — air-lock

Symptom
Erratic flow from tap, often with hissing and
spluttering

Cause
Air-lock (trapped bubbles)

Tools and materials needed
Hosepipe and tap-connectors or clips

Methods
1 Connect one end of the hose to a cold tap served
  direct from the rising main (usually, only the
  kitchen tap) and the other to the affected tap. If
  the kitchen cold tap is inconvenient, connect to any
  other tap. Turn both full on, and the pressure of
  the water should blow the bubble out. It helps to
  turn on other taps that are on the pipe-run that has
  the blockage. If it is a hot pipe, it helps to stop up
  the overflow pipe that goes from the hot system
  into the cold tank. (If airlocks occur often, a
  plumber can fit an air-release lock.)
2 Alternatively (unless the air lock is in a hot pipe),
  push a piece of hosepipe into the outlet of the cold
  tank and, with all cold taps turned on (except the
  kitchen one), blow down the hose.
3 If neither of these methods will do, drain the water
  system (see page 12). Then close each tap two-
  thirds before turning the water-supply on again.
  When all are flowing gently and equally, turn each
  tap on a little more (lowest ones first), then a little
  more again. Turn off equally gradually and in the
  same order.

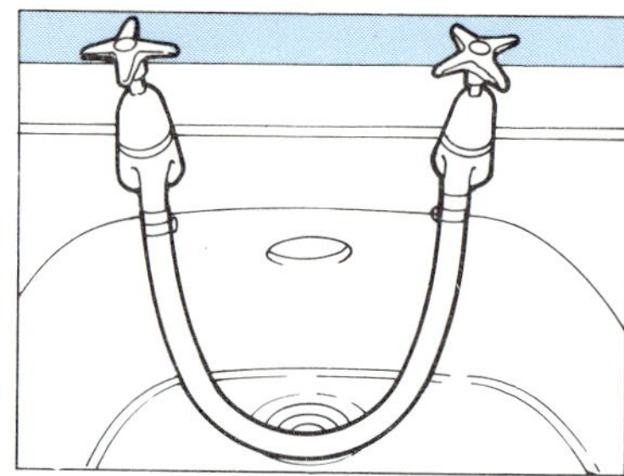

"

Plumbing-in appliances

Plumbing-in washing machines and dishwashers

Tools and materials
Plumbing kit (e.g. Siroflex)
Gimlet
Hand drill
Screwdriver
2 wallplugs (or epoxy putty)
Hose-connector
(Note: usable only on standard copper pipe, cold or hot)

Method
1 Screw wall-plate to wall behind pipe after drilling and plugging two holes in the wall.
2 Having drained water supply (see page 12), drill a hole in the pipe ($\frac{1}{4}$in (6mm) for cold, $\frac{3}{8}$in (10mm) for hot) starting the hole with a gimlet.
3 Place rubber seal round barrel before pushing it into the hole.
4 Push saddle-piece round the pipe and, with washers in place, screw to the wallplate. It is best to turn each screw a little alternately.
5 Screw L-shaped piece to the barrel.
6 With key provided, screw valve to L-piece.
7 Fix connector to hose. Restore water supply.

Extra connectors can be bought so that hoses serving two or three appliances can be attached to the supply as needed.

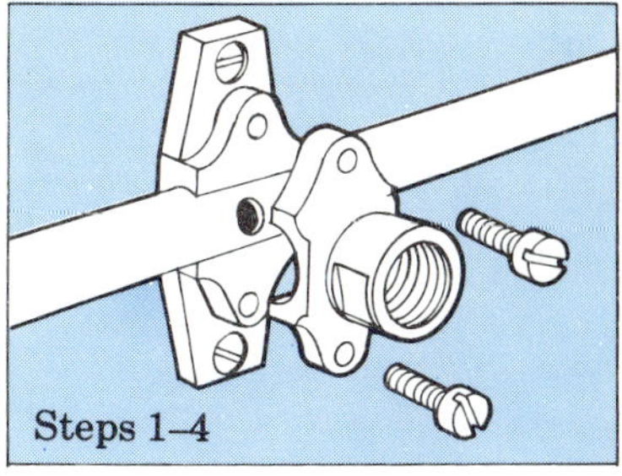
Steps 1–4

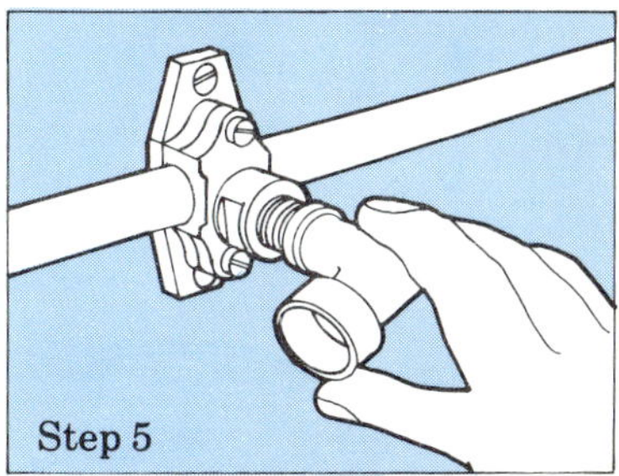
Step 5

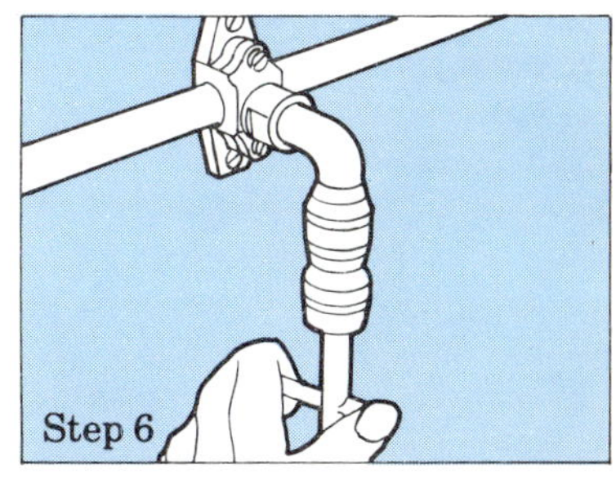
Step 6

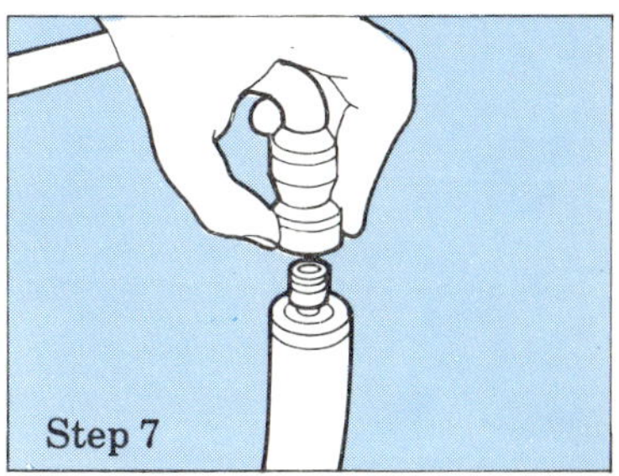
Step 7

Plumbing-in appliances

For machines not plumbed-in

A variety of tap connectors is available, many unsatisfactory. The diagram shows a tap nozzle adaptor that fits any shape of tap, is easy to tighten securely on the tap, has a threaded nozzle to fit straight into the hose without need for a connector and is rustless. Made in 3 sizes
The second diagram shows the special type of adaptor needed for a Supatap.
The third has an adjustable chain to hold the connector on, no matter how strong the water-pressure.
Where it is inconvenient to store a washing-machine or dishwasher alongside the water supply, it can be mounted on a special trolley (with brakes) and wheeled into position when wanted. A hose-clamp with suction base will secure the drain hose to the sink, preventing accidental floods.

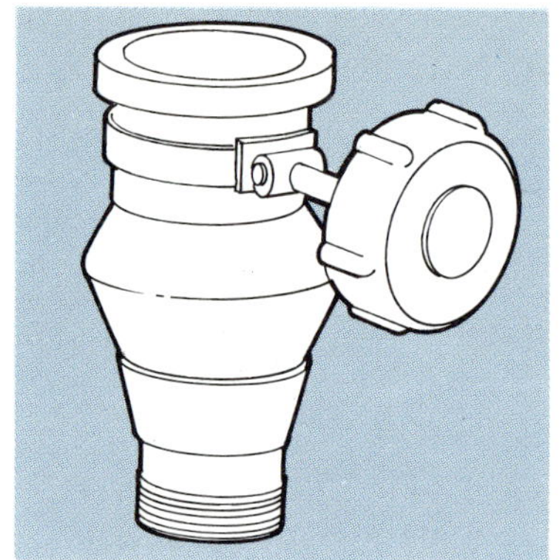

Scale-prevention

1 Ensure that the boiler or immersion heater is set to heat water only to 140°F (60°C), not hotter, if you live in a hard water area (see page 57).
2 Suspend a 'Micromet' descaling unit in the cold-water tank (this will need renewal about twice a year); where a cold tank is so close to a hot one that the cold water is in fact tepid, use 'Micromet X'. Or have an Albright scale reducing unit attached to a cold pipe (its crystals will need renewing periodically).
3 Alternatively, have a water softener installed (this not only stops scale forming in pipes but makes the water itself soft).
For advice on other hard water and scale problems, write to Albright & Wilson, Box 80, Oldbury, B69 4LN.

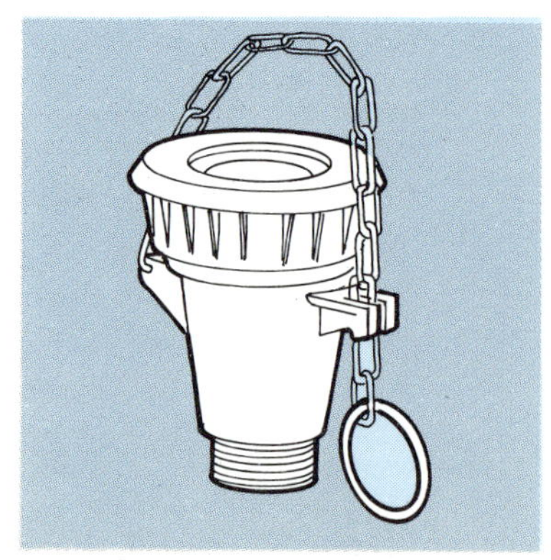

Water tanks — rust

Symptom

Rust traces (in the water supply or on the tank)

Cause

Corrosion of the galvanized steel tank

Tools and materials needed

Wire brush or abrasive paper
Goggles (or eye protectors sold by opticians, to clip
on spectacles) if using wire brush
Rust-killer (eg Jenolite) and steel wool
Possibly epoxy filler or plastic steel
Odourless bituminous paint (eg Aquaseal 44) or
zinc paint (these are suitable only for cold water
tanks, not hot)
Paint brush (or soft broom)
File (if zinc paint is used)
Brush cleaner or white spirit

Method

1 Drain tank (see page 12).
2 Dry it, and remove all loose rust with a wire brush
  or abrasive paper (wear goggles if you use a wire
  brush).
3 Coat with rust-killer, rubbed on with a steel wool
  pad and left to soak for 10 minutes.
4 If the tank is deeply pitted, apply epoxy resin (see
  page 32).
5 Apply two coats of bituminous paint.
  Alternatively, use zinc paint. The latter involves
  filing bare some patches of the steel, dabbing the
  paint on and, after 10 minutes, painting it
  liberally everywhere. After $\frac{3}{4}$ hour, splash water
  lightly all over.
6 Restore the water supply when the paint is dry,
  about 1 to 2 days.
  (An alternative is to line the tank with a flexible
  plastic liner: details from Plastic Liners Ltd.,
  Design Works, Three Colt St, London E14 8HH.)

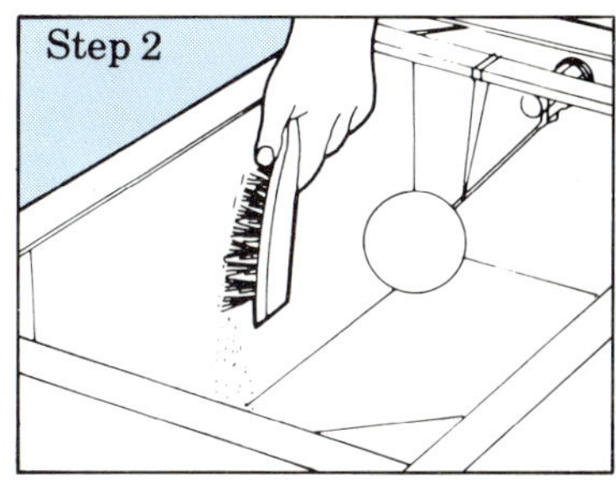

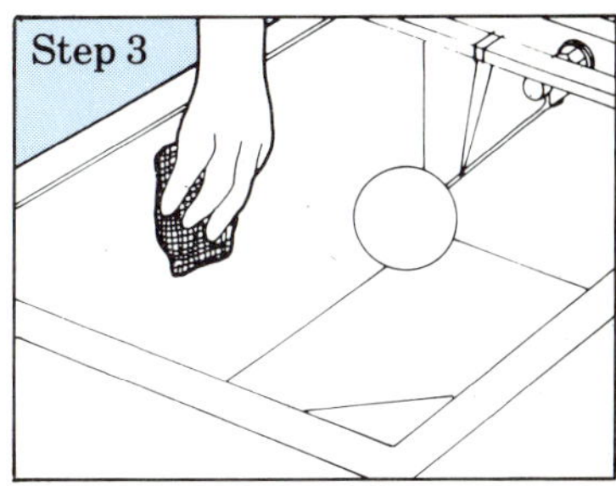

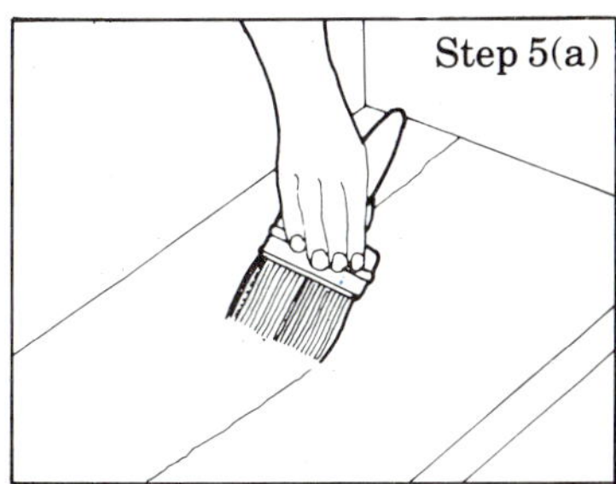

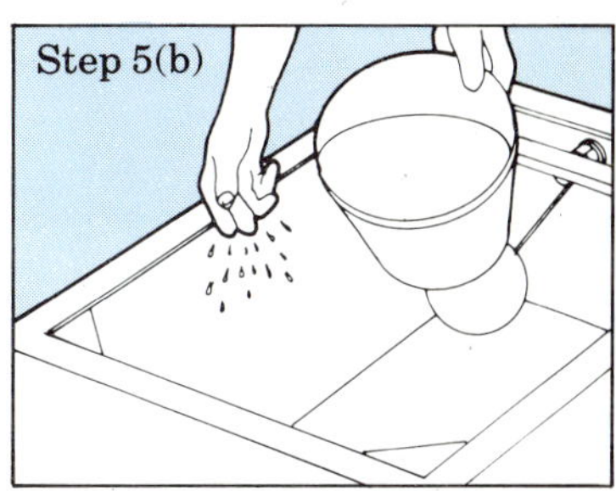

Water tanks — noise

Rust-prevention

Hang a 'sacrificial anode' low in the middle of the water (but not touching the ball-float). This is a block of magnesium which, gradually dissolving, has a protective action on the zinc coating of the galvanized steel tank. The block should be hung from a piece of wood by a length of copper wire, which needs to be secured to the tank edge by a small clip. File the place where it is to meet the edge, so that the copper wire can touch the bare steel. The anode will need to be replaced after a few years. Alternatively, 'Micromet' or an Albright-scale reducer (see page 64) will reduce corrosion.

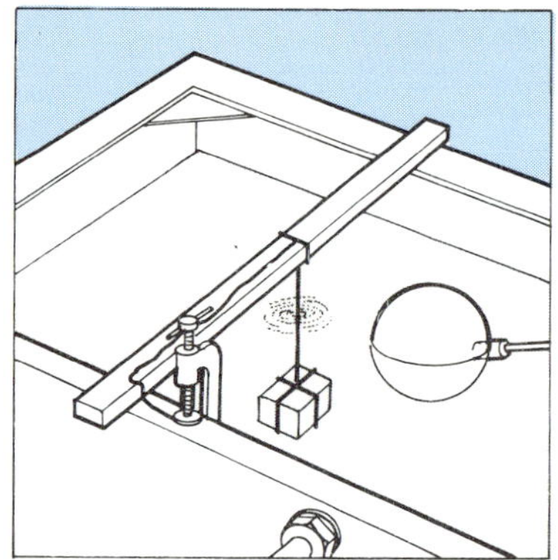

Symptom

Hammering or humming noise in the water system (see also Water Pipes)

Cause

Possibly, ball-float oscillating

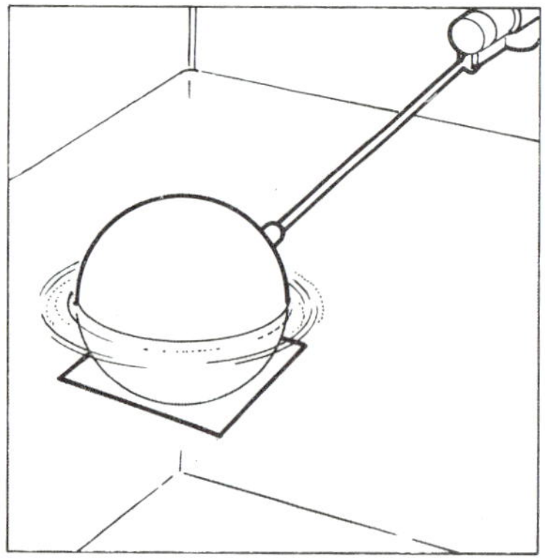

Materials needed

Larger ball-float
Alternatively, sheet copper and tube of liquid solder or epoxy adhesive
Or a plastic flowerpot

Method

Unscrew the old float and replace with new one. Alternatively, if the float is a metal one, fasten a piece of sheet copper to the bottom of the existing float in order to steady it; or, if it is a plastic float, hang a plastic flowerpot (right way up) from the arm so that it hangs just below the ball float. (If this fails, a plumber can fit an air chamber to the cold tank.)

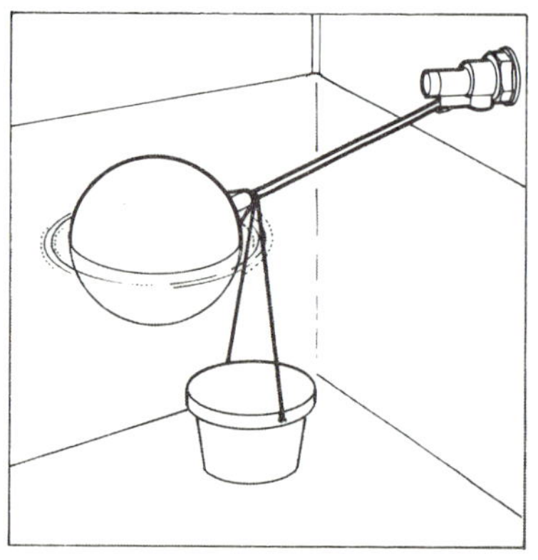

Water tanks — overflowing

Symptom
Drips or stream of water from overflow pipe

Cause
Faulty ball-float; or deteriorated ball-valve washer

Tools and materials needed
Possibly new ball float (plastic)
Pliers
Possibly new washer, screwdriver, Vaseline, and
steel-wool for cleaning

Method
1 Unscrew the float and shake: if you hear water
inside it is leaking and needs replacement. To keep
water from entering the tank or cistern until the
new float is screwed onto the arm, tie the arm to a
piece of wood lying across the top of the tank as
shown on page 42. Alternatively, enlarge the hole,
shake the water out and – after screwing the float
back – tie a plastic bag round it.) If there is no leak,
adjust the float arm: bending it down an inch or so
(gently) will shut off the water-supply at a lower
level. (a few have an adjustable nut on them).

2 If overflow continues, draw off water supply (see
page 12) and use pliers to remove the splitpin
holding the arm. If the pin looks worn, replace it.
In the case of a WC cistern, you may have to
remove the flush-handle first in order to get at it.
Coax the arm out of the slot in the valve.

3 Unscrew the cap at the end, if there is one, and
slide the plug out with the help of a screwdriver if
necessary (there is a slot in it for this purpose).
Tapping gently round the end with a hammer may
help.

4 Unscrew the two halves (you may need to apply hot
water or a release lubricant) and put a new washer
in. (Rather than damage the plug while forcing it
apart, it may be better to pick out the old washer
with a penknife and squeeze the new one in.)

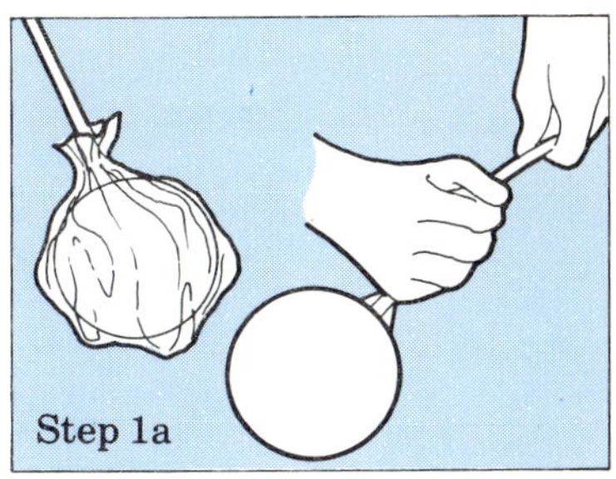

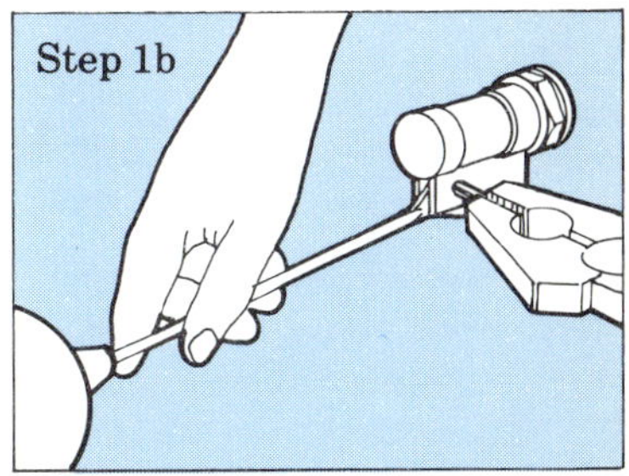

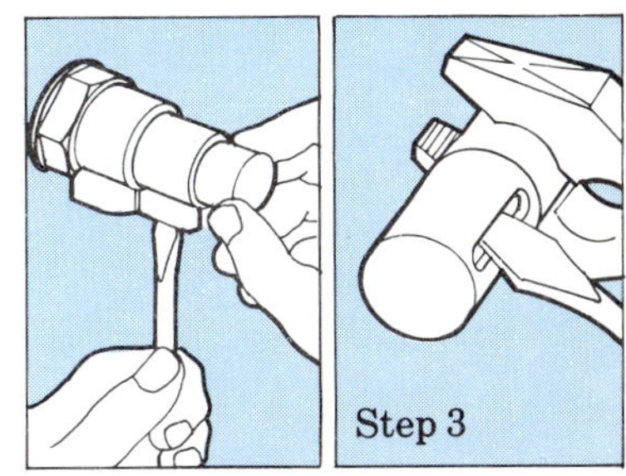

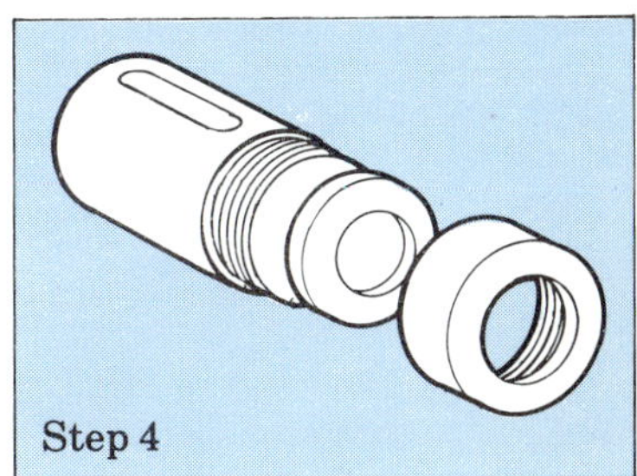

Water tanks — slow refilling

5 Clean and grease the plug and the inside of the valve, before re-assembly.

Symptom
Slow refilling

Cause
Possibly, scale or grit in ball valve

Tools and Materials needed
Pliers,
screwdriver
Vaseline

Method
As for the rewashering ball-valve (see page 43). (If this is ineffective, plumber may need to put in a different type of valve.)

Note: Some old-fashioned ball valves can be replaced completely with diaphragm valves, less likely to go wrong (a job for a plumber).

Cisterns and WCs — noise

Symptom
Noisy flushing

Cause
Old valve

Tools and materials needed
Torbeck float valve
spanner
scissors

Method
1 Having drained off the water (see page 12) and
unscrewed the old valve, pass the connector of the
new valve through the hole in the cistern, in order
to hand-screw it to the inlet pipe (with washer
between pipe and cistern). Check that the
polythene tube of the valve is hanging straight
down, then tighten the nut with a spanner.
2 Snap the rod of the float onto the float-arm. (The
rod length can be adjusted after the water is
restored, and any surplus cut off.)
3 Restore the water supply and check whether the
valve operates properly when the WC is flushed.
(If, after repeated operation, it fails to close
properly, the cause is likely to be debris from the
cold-water system. Open up the valve and clean it
out.)

Other noise-reduction methods
Put rubber sleeves round screws fixing WC to wall.
Put resilient draught strips round and under WC
door, and panel both sides of it.
Box in waste-pipes with thick chipboard, sealing
all gaps, but leave space between boards and pipes.
Pipes should not fit too tight in clips or holes; nor
be so loose that they vibrate against hard surfaces
(use resilient padding, polyurethane foam, or
something similar to cushion them).

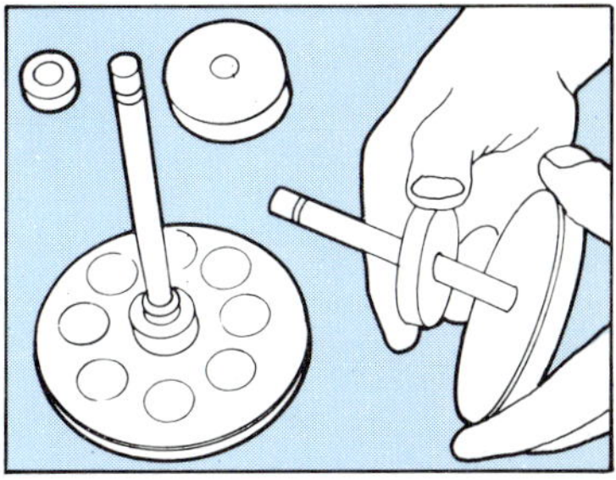

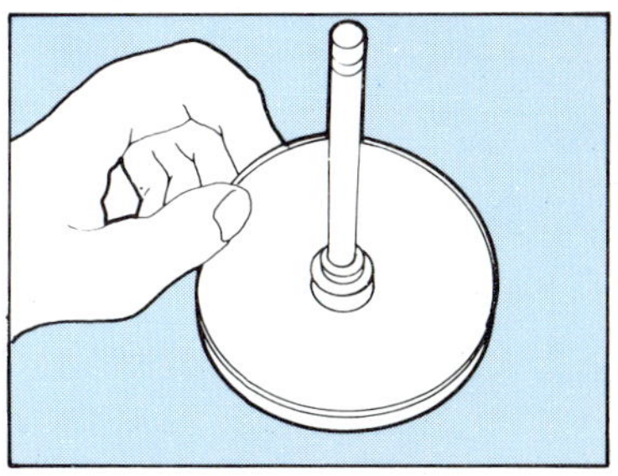

Cisterns and WCs — noise

Symptom
Gushing noise in cistern

Cause
Water filling the cistern after flushing

Materials needed
Short tube with screw end
Vaseline

Method
1 Tie up the arm of the ball-float (see page 42) and flush the WC to empty the cistern.
2 Grease the screw-thread and screw it into the hole that is beneath the ball-valve.
3 Release the arm.

Other noise may be remedied by the flowerpot method described on page 42.

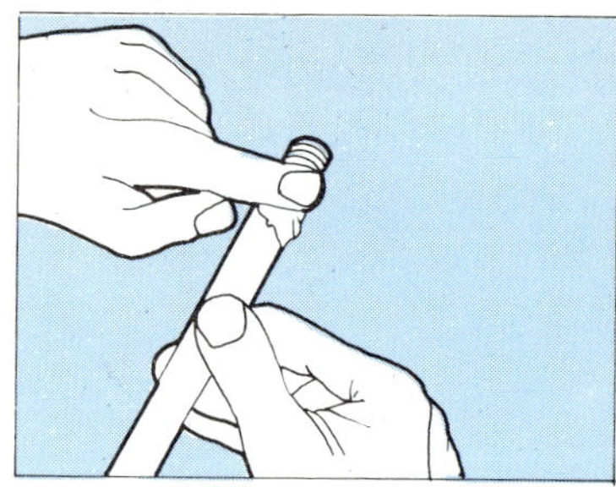

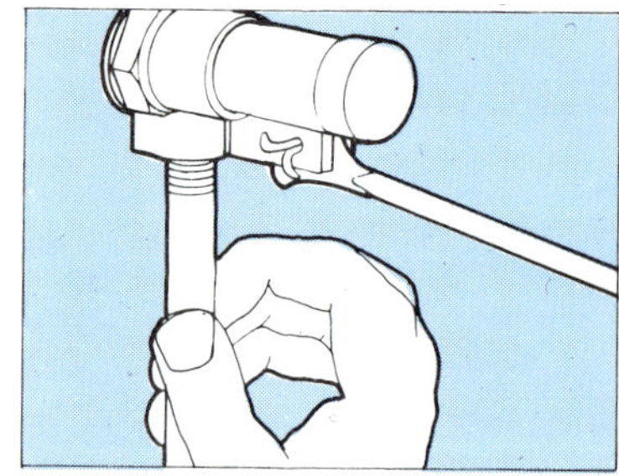

Step 2

Cisterns and WCs — failure to flush

Symptom
Failure to flush without repeated jerks of the
handle

Cause
Water-level too low; or deteriorated valve in the
siphon

Tools and materials needed
Spanner
Siphon flap-valve

Method
1 If the water-level is more than 10mm below the
  overflow pipe, detach and bend the arm of the ball-
  float up a little.
2 If the water-level is correct, tie up the float arm
  (see page 42) and flush the cistern to empty it.
3 Disconnect the flush pipe, unscrew the nut below
  the cistern, and lift out the siphon.
4 Replace old valve as shown in diagram.
5 Reassemble and release float arm.

Cisterns and WCs — condensation

Symptom
Drips from exterior of cistern

Cause
Condensation due to inadequate ventilation in the
room

Tools and materials needed
Anti-condensation paint, paintbrush and brush
cleaner (or white spirit)
Or Spontex foam strips (as sold for condensation
that runs down windows and on to sills)
Or Polystyrene (as sold for lining walls),
sandpaper, scissors and waterproof adhesive

Method
1 **Anti-condensation paint** (eg Korkon)
  (Note: this is suitable only for iron cisterns not
  ceramic ones. It is often sold by ship's chandlers.)
  Paint on in the usual way then finish with a
  stippling action. It has a textured look because it
  contains cork granules. After a day, it can be
  painted (its natural colour is off-white).
2 **Foam strips** Stick round bottom of cistern to catch
  the drips.
3 **Polystyrene** Drain cistern (see page 46) and after
  sandpapering inside it glue polystyrene on. Dry
  overnight.

WC CISTERN

Symptom
Overflowing
Slow refilling (see Water tank, page 43)

Descaling acid ('Descalite') or WC cleaner are
useful for dealing with the crust that sometimes
forms under the rim of WC's. To make it adhere,
mix it to a paste with (for example) Polycell or
even flour and water. Leave for $\frac{1}{2}$ hour before
flushing away. Some WC cleaners are harmful.
WC manufacturers recommend Harpic and Dot.

Cisterns and WCs — uneven flush

Symptom
Uneven flush in bowl

Cause
Obstruction; or tilting of pan

Tools and materials needed
Spirit level
Screwdriver
Dryish mortar made with one part cement to 3 of
sand (or buy ready-mixed mortar)
Grommets

Method
1 Check that nothing is obstructing water inlet or
  rim (a mirror will help you see under this).
2 If the spirit-level shows the pan is not level, loosen
  the screws fixing it to the floor. Remove those on
  the side that is to be raised.
3 Raise this side with scraps of board until the spirit-
  level shows the pan is straight.
4 Pack mortar below. Replace screws loosely, with
  grommets (sleeves) round them, and screw tight
  three hours later.
  Warning: do not attempt this if pan is rigidly
  fixed to an iron waste-pipe; you might crack the
  join.

Note
Before buying a spirit level, check in the shop that
it is 'true'. Use it on a level board across the WC.
Alternatively, use a glass dish of water, marking
the water level (when on a true surface) with
sticky-tape.

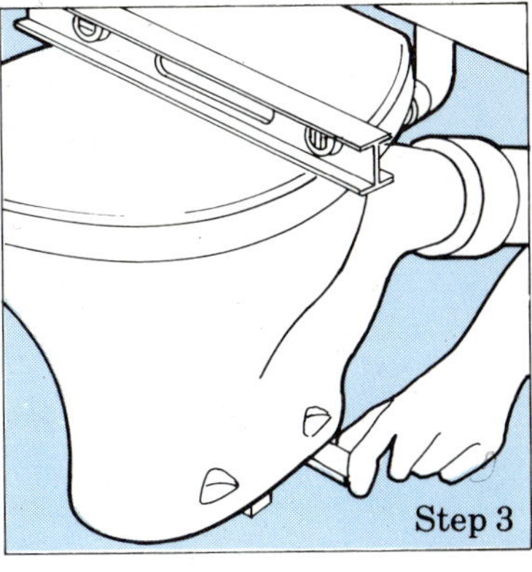

Step 3

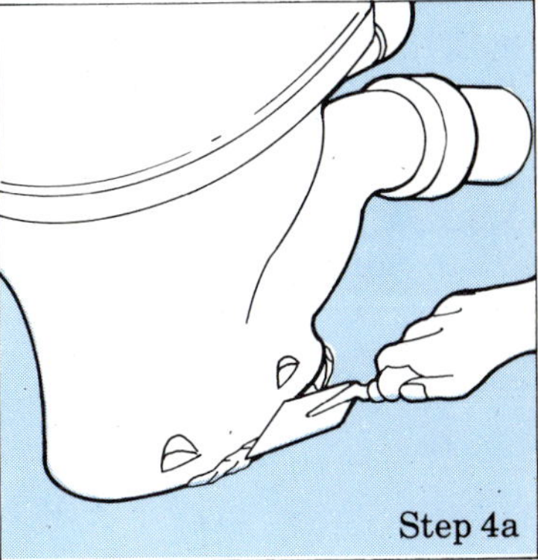

Step 4a

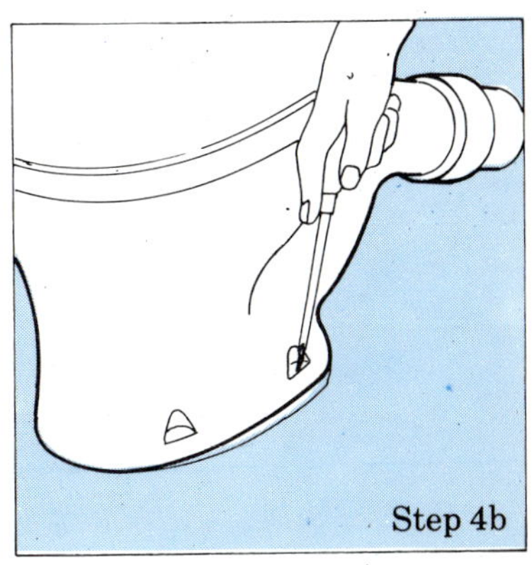

Step 4b

WC seat and lid replacement

Seat and lid replacement

Tools and materials needed
New seat and lid set (tell retailer the size of the
bolt holes in the WC, and their distance apart)
Pliers
Screwdriver

Method
Undo nuts fastening old seat to WC, and remove it.
Follow manufacturer's instructions for assembling
and fixing new set. Do not over-tighten wing nuts.
Note
A break in a plastic seat or its hinge can be
repaired with epoxy putty (see page 20).

Water shortage

In a drought or a freeze, water may be too precious to waste. To reduce the amount lost in flushing a WC, put into the cistern a plastic bag of water (fastened with a rubber band).

To siphon bathwater out to re-use for household cleaning, clothes washing or gardening, two people are needed.

Method

1 With one end of a hose (A) attached to tap, raise other end (B) above tap level and cover it with a thumb while the hose is filling with water from the tap.
2 Detach end A and raise it above end B.
3 With thumb still covering end B, lower it below the water in the bath.
4 With thumb now covering end A, lower this to any receptacle below bath level and let go. The bath water will now flow into the receptacle.

If there is an accessible down-pipe outside, a down-pipe adaptor can be bought and fitted to it: this enables bath or washing-machine water to be fed to a garden hose. Soap and detergent benefit plants, but do not use water that has chlorine bleach in it.

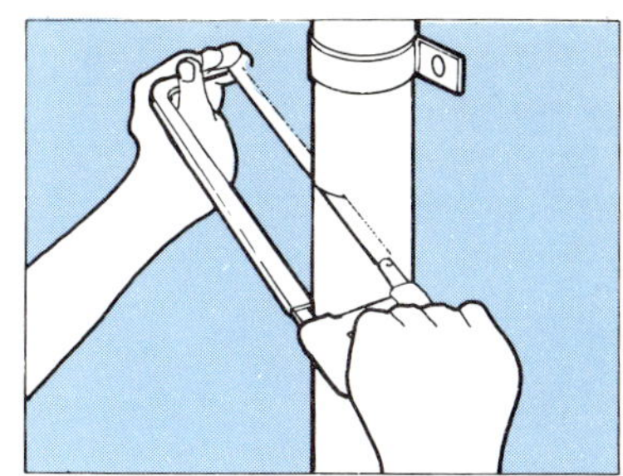

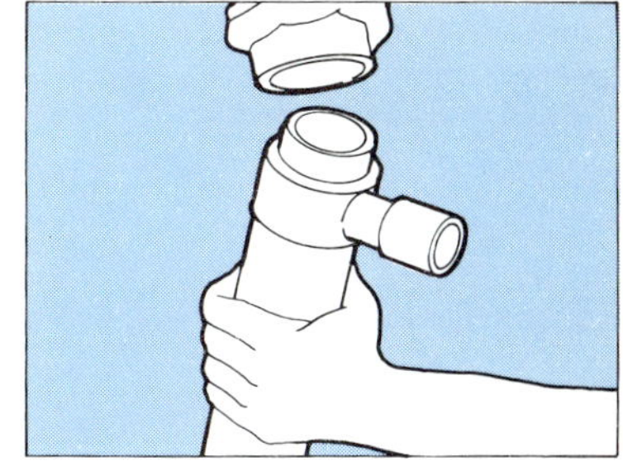

a down-pipe adaptor fitted to a water-butt

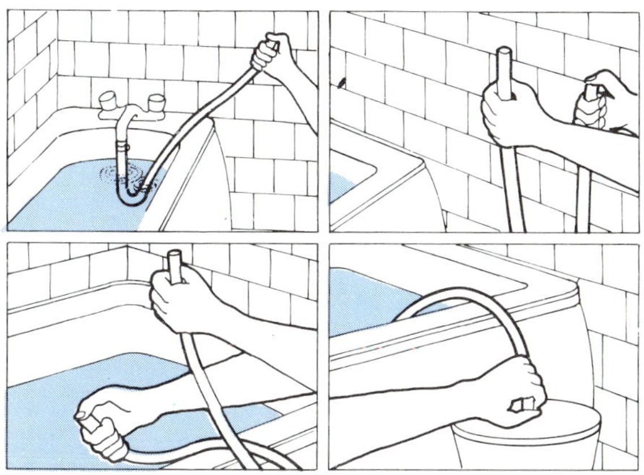

Bathroom fitments — maintenance

Replacement of plug chain in basins and baths

Materials and tools needed
Chain
Pliers

Method
Push bolt on end of chain through hole in bath, and fasten nut and washer at the back. Use the pliers to open the ring on the other end and fasten it through the loop on the plug.

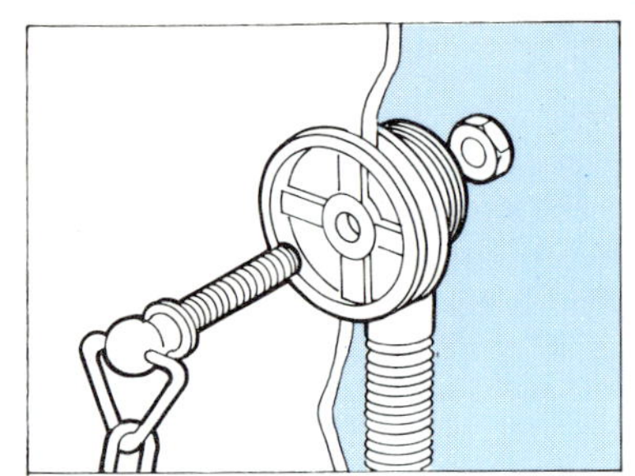

Cracks, chips
(in basins, baths, sinks, WCs)

Temporary repairs can be made with waterproof sticky-tape on the underside of cracks; or cracks and chips may be permanently filled with epoxy putty (see page 20). Allow time for drying before running any water on. Use Joy bath enamel or 'Porcelainit' to paint the repair (several coats). Some bath manufacturers supply their own filler for repairs.

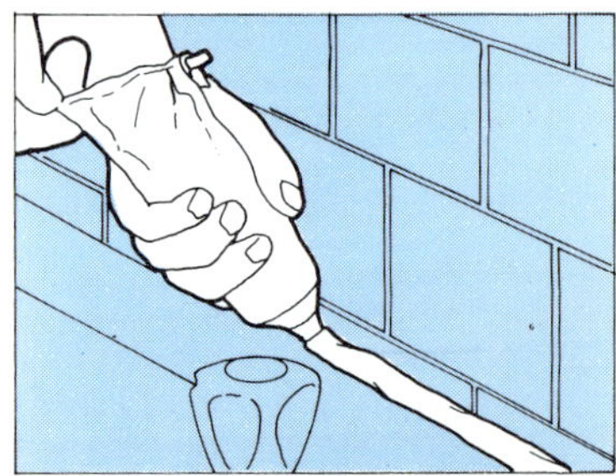

Gaps at wall
(round basins, baths and sinks)

As a hard-setting filler would probably crack, use a flexible and waterproof bath sealant, preferably made of silicone rubber. Clean off all soap traces first, and squeeze the sealant well inside the gap. Alternatively, to fix outside the gap, use Seel-a-Round plastic tape supplied with contact adhesive, or press-in sealing strip (see page 21).

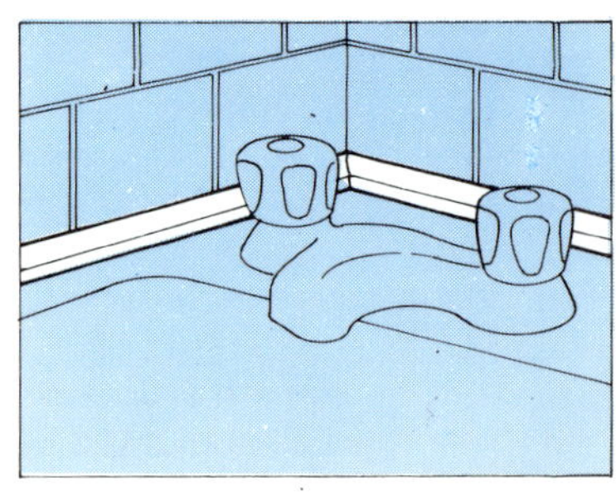

Bathroom fitments — maintenance

Maintenance

Dinginess of bath and basins may be due to hard
water, soapy deposits, rust from the water
(particularly if a tap keeps dripping), coppery
stains from an old gas water-heater or a
combination of these.

Daily care

A tablespoon or two of Calgon water-softener in
bath or washing water will prevent the build-up of
hard water and soap deposits; or a liquid detergent
like Stergene will help if used whenever bathing or
washing. Wiping the bath or basin over while still
warm is more effective than cleaning later;
abrasive scouring powders can do actual harm on
some enamels, so use only products labelled as
being intended for baths – whether they are
powder, paste or liquid. Ceramic washbasins or
vitreous enamelled surfaces, however, will not be
harmed by ordinary cleansers. Paint cleaners like
'Polywash' are very effective bath cleaners; so are
paint brush cleaners like 'Polyclens'.

Renovation

Turpentine or methylated spirit may shift soapy
deposits (and also hair lacquer on mirrors).
'Renubath RB70' is an acid cleanser that can be
used to deal with heavier deposits but may
diminish the glossy look of enamel, so it is used
diluted on enamel but full-strength on ceramic
washbasins, etc., which it will not harm. 'Jenolite'
is an acid jelly that deals with rust stains as well
as hard water marks.

Perspex baths sometimes get slightly scratched:
rubbing with metal polish may improve matters.
For deep scratches or cigarette burns, rub gently
with steel wool first and then use metal polish.

Hard water deposits around the base of taps may
be shifted with the help of an old toothbrush and
bath cleanser, but if this does not work a descaling
acid (fluid or powder) could be used. You may find

Bathroom fitments — maintenance

the enamel is in poor condition under the hard water scale after you have removed it.

To get at the deposits inside washbasin overflows, use a wire-handled bottle brush which can be bent to fit.

Hydrochloric acid is usable for stains associated with hard water but is dangerous stuff – both poisonous and corrosive – so it is wise to wear rubber gloves, keep it away from children, and flush away any leftovers rather than have a risk in the house. It can attack some enamels, and also metals (coat wastepipe holes with Vaseline if using the acid near them). Rinse away the minute the desired effect has been achieved.

Coppery marks (green) may yield to an acid: vinegar will not harm enamel, but anything stronger may do so if the surface is not vitreous enamel (glass-like) but only enamel paint.

Special bath enamels are sold for repainting baths, but a better job will be done by getting in a professional firm who will first repair chips and scratches, using infra-red rays, clean off discolorations with powerful chemical polishes, and spray on 'Vitrocoat', an epoxy resin. The bath will be out of action for one week while the new surface hardens. (Details from Renubath, 596 Chiswick High Road, W4.) If you do decide to re-enamel a bath yourself, use special epoxy bath paint.

It is good practice to run cold water first, then hot (with bath salts last) as this is kinder to the enamel and also helps to reduce condensation.

Chromium accessories may show traces of rust. Of the many rust removers available, 'Plus Gas' or 'WD40' (in aerosols) are easiest to use, penetrate even small gaps, and not only remove rust but provide some future protection. Do not use abrasives on chromium or other plating. Special chromium polishes are sold but should not be necessary.

Bathroom fitments — bath

Insulating a bath (or steel sink)
To conserve the heat of the water, a coating of
polyurethane foam can be applied to the underside.
It is usually easier to put the foam onto sheets of
wrapping paper and then press these onto the
metal while the foam is sticky.

Either : Brush water lightly over the paper, then
brush the foam evenly all over it, working it into
the water.
Or: Paint the foam on then spray a little water
over it.
As the foam sets it expands. Another layer can
then be applied if a thicker jacket is wanted.

Showers

Symptom
Reduced flow

Cause
Scale in shower-head or mixing valve

Tools and materials needed
'Scale-Away' descaling powder
Or 'Descalite' liquid
Or vinegar

Method
Unscrew affected parts and soak for an hour in $2\frac{1}{4}$
litres of nearly boiling water to which the powder
has been slowly added, or in 'Descalite' or vinegar.

Bathroom fitments — showers

Symptom
Continual dripping

Cause
Deteriorated O-ring (a kind of washer)

Tools and materials needed
Screwdriver
O-ring

Method
1 Having turned the water supply off (see page 12) unscrew the flexible hose.
2 Unscrew the shower-diverter.
3 Beneath this is a connector with a slot: holding screwdriver in this slot as shown, push the connector off.
4 Pull out the mechanism from the body of the tap, slide off the old O-ring, and put a new one on.

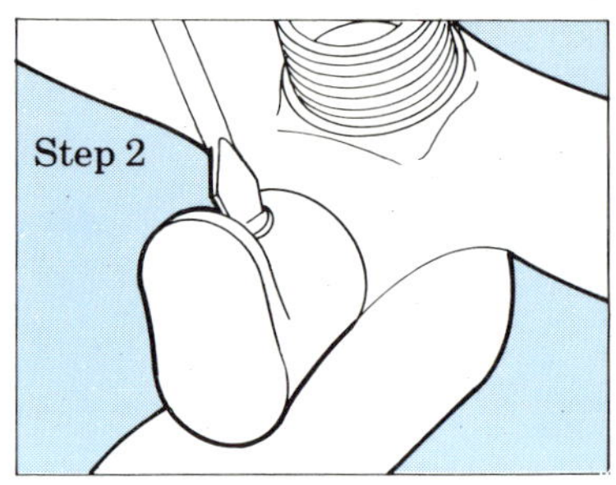

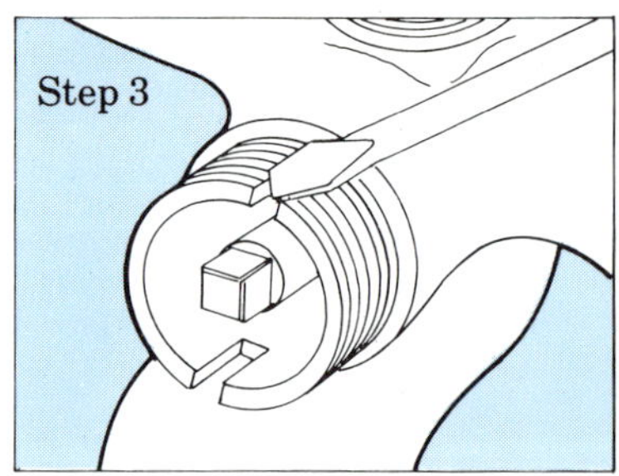

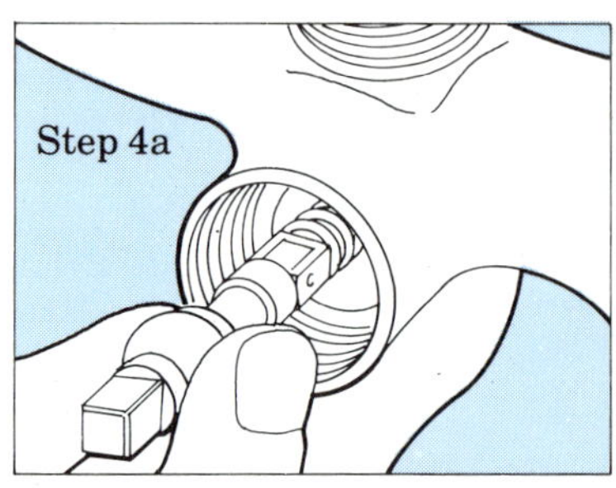

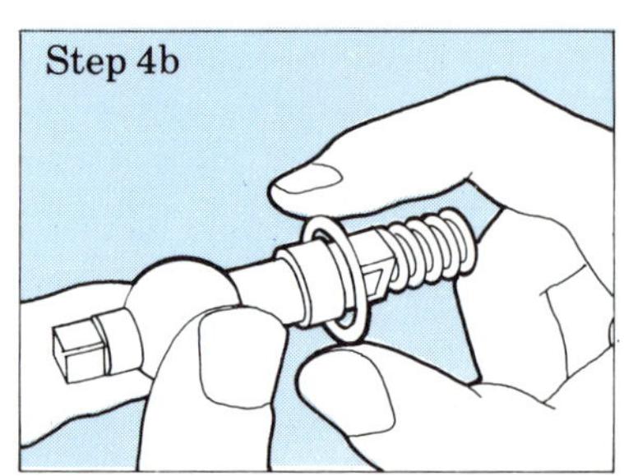

Water heaters — electric immersion

Symptom
Water too hot or cold; slow reheating

Cause
Thermostat wrongly set
possibly scale from hard water

Tool needed
Screwdriver

Method
1 Turn off electricity.
2 Undo screw holding the cover on the element.
3 Prize off cap covering the regulator.
4 Use screwdriver to turn screw that regulates the setting. 140°F (60°C) is normal in hard water areas, 160°F (70°C) in soft water areas.

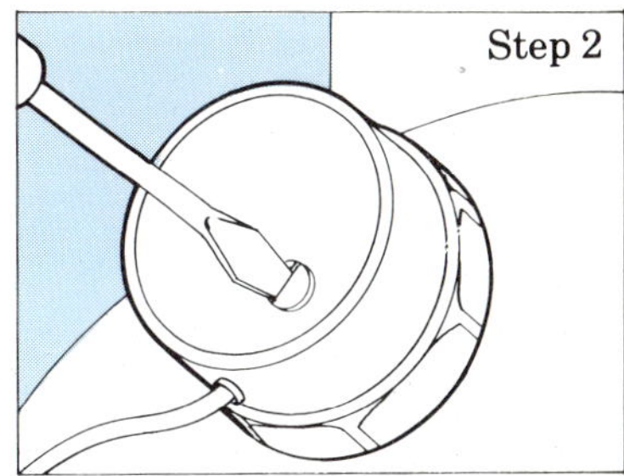

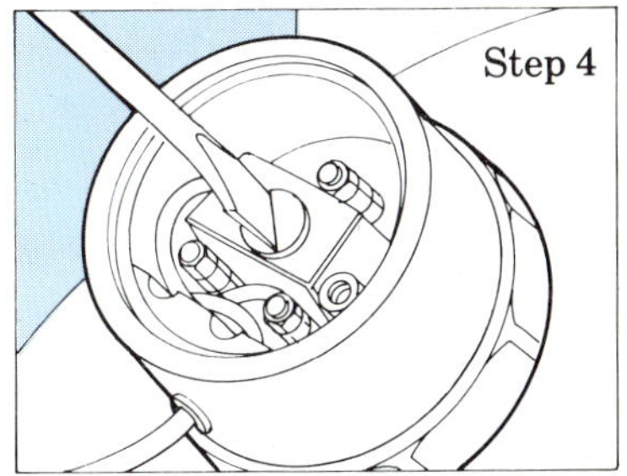

Hot-water tanks
Prevention of heat-loss
All hot-water tanks, whatever their heat source, should be lagged. Buy a 3in (75mm) thick padded jacket to fit exactly. Instructions will be included.

Hot-water pipes can be lagged in the same way as cold ones (see page 35). The vent pipe rising from the cylinder should be lagged too if it rises vertically (it ought to start horizontal for about half a metre).

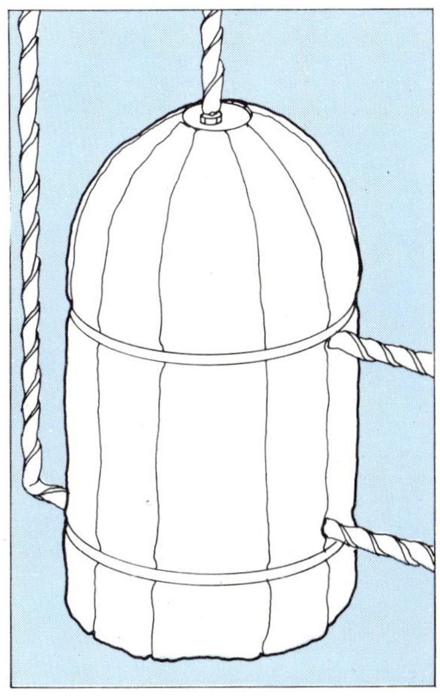

Water heaters — gas

Symptom
Pilot flame inadequate

Cause
Wrongly adjusted, or clogged

Tools and materials needed
Small screwdriver
Primus pricker (sold by gas showrooms)

Method
1 Remove front of casing (pull off control knobs first if necessary).
2 Use screwdriver to adjust the screw on the pilot while it is alight.
3 If the jet is blocked, turn pilot off and use the pricker to clear it.
4 Reassemble and, after lighting pilot again, wait 5 minutes before using the water heater.

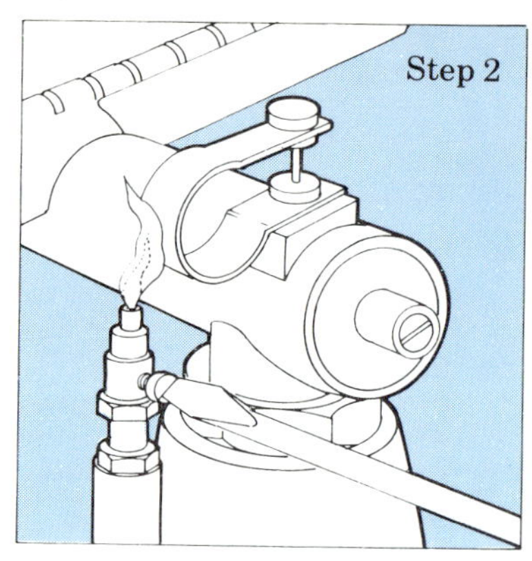

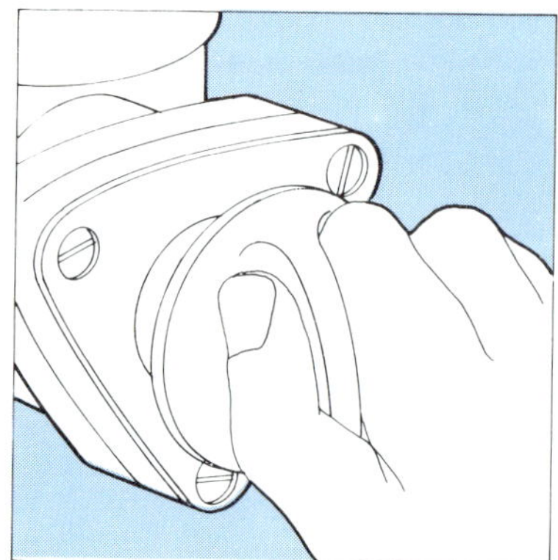

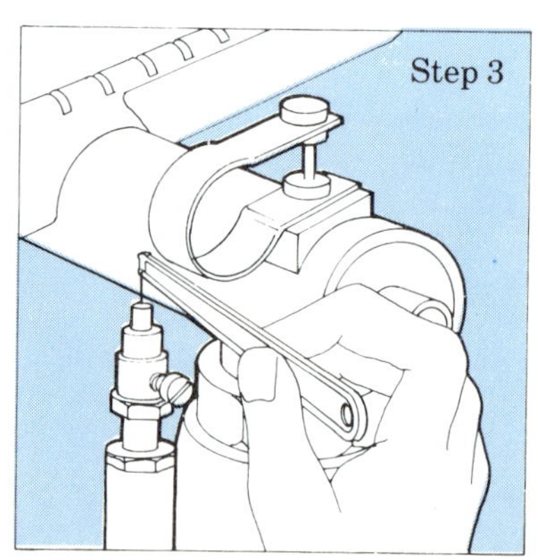

Water heaters — gas

Symptoms
Poor flow or not hot enough; drips from spout

Cause
Possibly, scale from hard water (removing this in a *large* heater is a job for a gas serviceman)

Tools and materials needed
Rubber tube
Plastic funnel
Descaling fluid

Method
1 Turn off water and gas supplies (see pages 12 and 63).
2 Disconnect water inlet, and in its place attach about a metre of rubber tube.
3 With end of tube held above top of heater, pour fluid into it through the funnel – very slowly to prevent it foaming back.

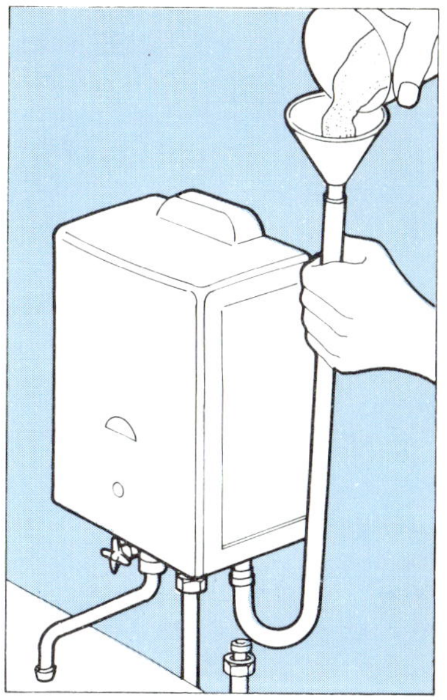

Symptom
Burners fail to light when water is turned on, or light feebly
Cause
Possibly, metal strip by pilot light is not functioning; or gas filter needs cleaning; or gas pressure needs increasing (see also Pilot flame, page 58)

Tools and Materials needed
Screwdriver
Possibly new bimetallic strip

Methods
Check that control knob or lever is fully in 'on' position; and that gas main tap is fully on.

Metal strip
1 Press strip down with screwdriver tip (with pilot light on). If burners now light, this means strip is

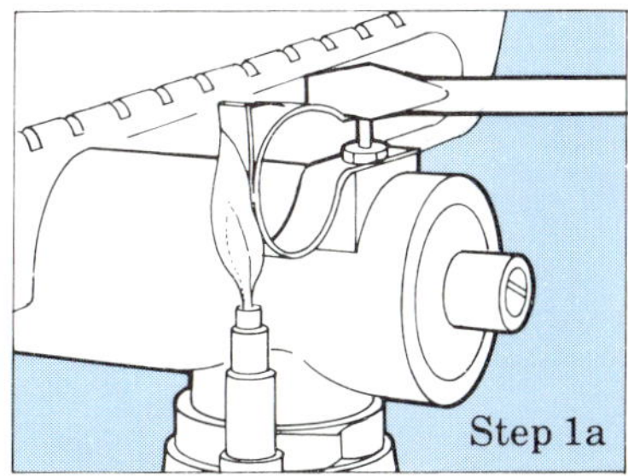

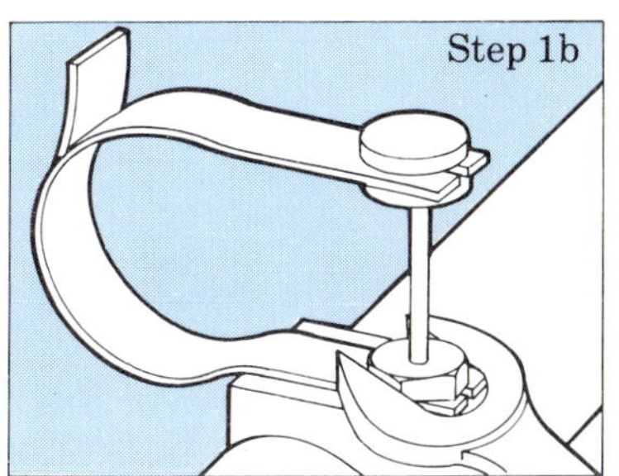

Water heaters — gas

at fault. Turn off gas and put in a replacement
(loosen the nut to free the old one from the pin).

2 Alternatively, if strip is not at fault, press pilot
fitment gently nearer to the strip so that its flame
can make contact.

3 If the pin holding the metal strip is too tight, turn
off gas and work it up and down repeatedly to
loosen it.

Filter

Look for this on gas inlet pipe. Undo cover; take
out and clean filter; reassemble.

Pressure

Look for governor on gas inlet pipe. Remove cover
and adjust screw till pressure increases. One cause
of low pressure can be rust flakes in old pipes: the
Gas authority can vacuum these out (no charge).

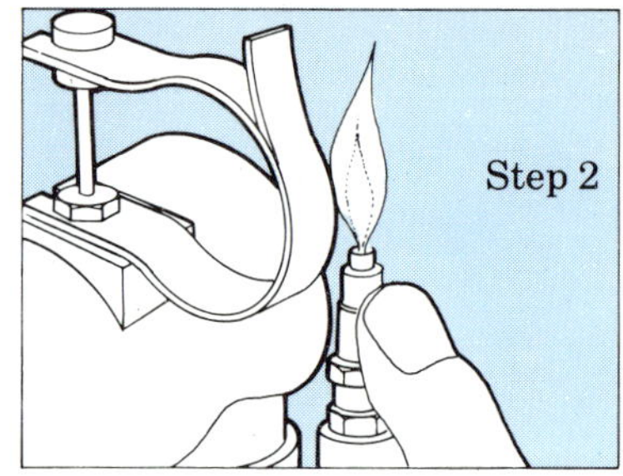

Step 2

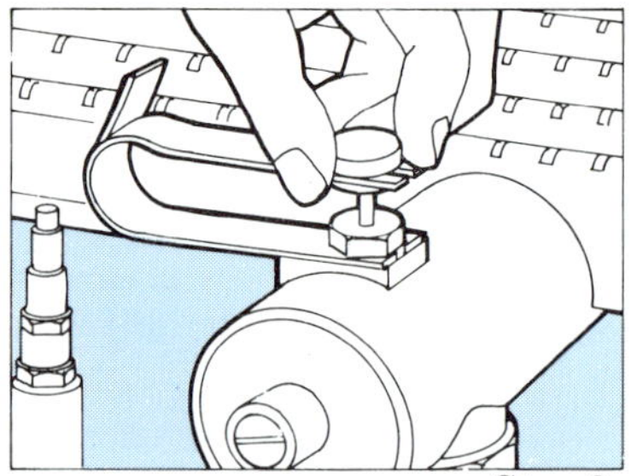

Step 3

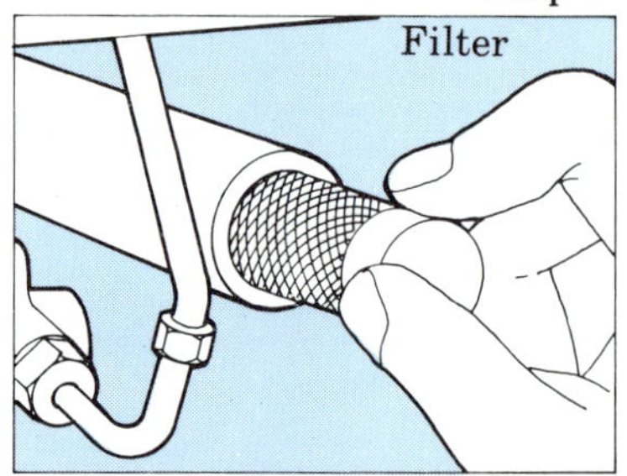

Filter

Pressure adjustment

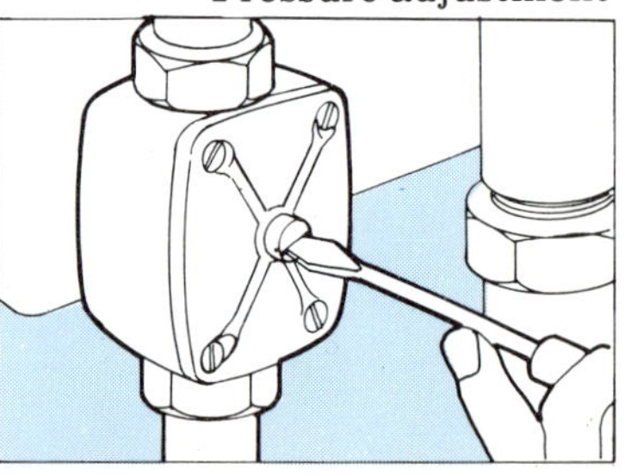

Water heaters — gas

Symptom
Lighting with sudden roar

Cause
Build-up of gas before pilot lights the burners

Tool needed
Screwdriver

Method
With gas off, gently bend pilot fitment nearer to burners.

Symptom
Leaks from joints

Cause
Loose nuts

Tools and materials needed
Spanner
Vaseline
Possibly washers

Method
Grease and then tighten (but not too much), replacing any worn washer.

Water heaters — gas

Symptom
Smell of burnt gas when in use (this is dangerous)

Cause
Blockage in ventilation duct (or running a small unflued heater for more than 10 minutes)

Method
1 Hold a match under the cowl while the water-heater is on. If this goes out, it may be because the burnt gas is coming back into the room.
2 Check from inside and outdoors whether duct is blocked by, for instance, a bird's nest which you can remove. If this is not the cause, call Gas Service.

Note
Any room containing a small unflued heater should be well ventilated. One simple method is to drill a row of holes through both a window-frame and a door to ensure a current of air, but window-ventilators or airbricks are neater.

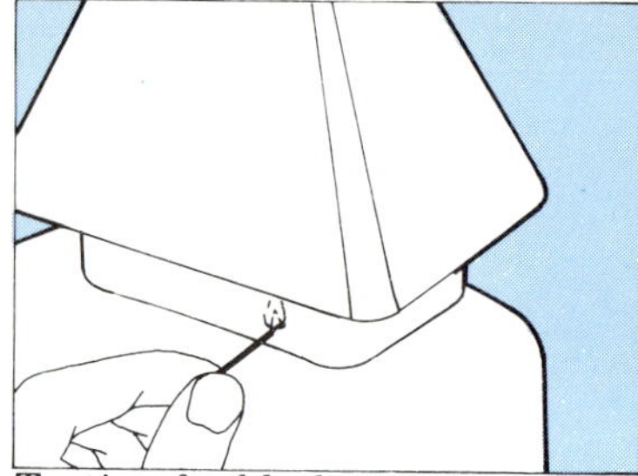
Testing for blockage

Ventilation holes

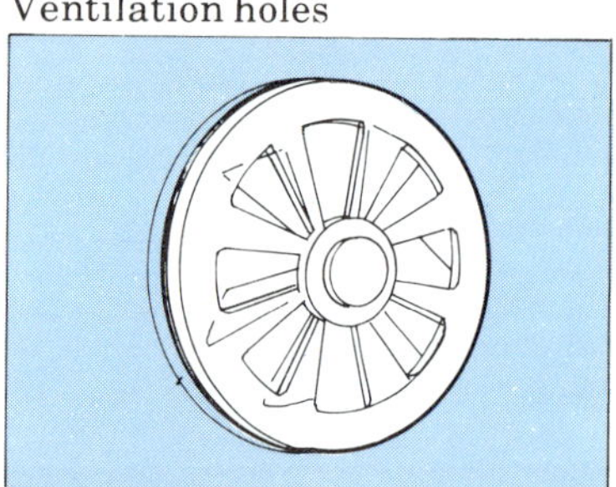
Window fan

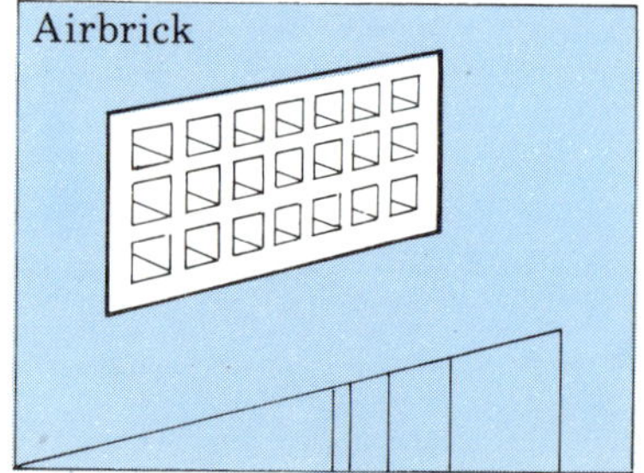
Airbrick

Water heaters — gas

Gas Emergencies

If there is a strong smell of gas
Turn main tap off; extinguish any flames or fires;
open windows.
Phone Gas Service: put their day and night
numbers here..

....

If there is a slight smell, owing to pilot light
having gone out:
Turn pilot off; extinguish any flames or fires; open
windows. When smell has gone, turn pilot on and
light it.

**If gas in water heater continues to burn after
water is turned off:** turn gas off for a while. If
this trouble recurs, phone Gas Service.

Boiler — descaling

Symptom
Poor flow of hot water; sometimes with gurgling, hissing and knocking sounds

Cause
Scale from hard water (you can check whether this is the cause by seeing whether your kettle, too, has scale in it)
Descaling pipes or a hot water cylinder is a job for a plumber; descaling a boiler can be done by an amateur

Tools and materials needed
Method A
Albright boiler descaler (6kg for an average boiler)
Hired descaling unit (Both these are obtained from R F Symons Ltd., 9 Shaftesbury Parade, South Harrow HA2 0AJ)
Hose and clips
Method B
Descalite boiler and central heating kit
Method C
Ferndox DS9 liquid

Method A
1 Turn off or rake boiler out. Drain hot water system (see page 12).
2 Check that the drain cock on the boiler has no leak (if it has, the washer must be renewed before proceeding further: for method, see page 24) and connect tube of descaling unit to it.
3 Pour into the unit 1kg of the descaling powder and add 2 pints (1 litre) of nearly boiling water; shake to mix.
4 With the unit standing 6 feet (2 metres) above the floor, turn its tap on so that the fluid flows into the boiler via the drain cock.
5 Repeat this operation until 3kg of the descaling powder has been used and then wait 3 hours or until the liquid is motionless and silent (it gurgles while it is active).

Step 2

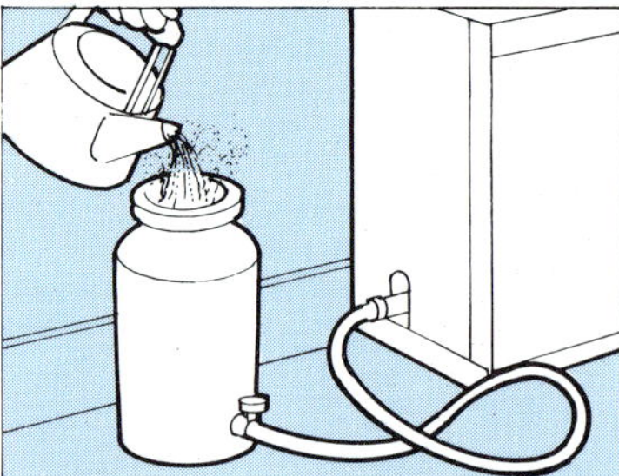
Step 3

Step 4

Boiler — descaling

6 Repeat steps 3 to 5 with the remainder of the
   powder.
7 Restore the water supply and flush the system
   clean for $\frac{1}{4}$ hour by means of a hose connected to the
   drain cock and leading to an outdoor gulley.
8 Turn off the drain cock and disconnect the hose.
   After refilling and lighting the boiler, run hot
   water from all taps until the water is free of all
   colour.

Method B

1 Switch off boiler and turn off water supply to
   boiler tank. Drain the boiler wearing rubber
   gloves.
2 Pour descaling liquid into the tank, then turn the
   water supply on again.
3 Switch boiler on again and run at low temperature
   for 3 hours.
4 Turn off water supply, drain boiler
5 Dissolve neutralising crystals in hot water and add
   to tank.
6 Turn on boiler for $\frac{1}{4}$ hour, drain, flush with clean
   water.
7 Close drain cock, fill with water.

Method C (central heating system)

1 Drain the hot water system.
2 Fully open radiator valves and vents.
3 Add the acid to the header tank while it refills
   with water (allow 1 gallon ($4\frac{1}{2}$ litres) of Fernox DS9
   for every 20 gallons of water in system and for
   every 5 years of its age).
4 With temperature control at 150°F (65°C) leave to
   circulate for 2 days. (Repeat steps 3–4 if system
   needs more than 1 gallon acid per 20 gallons
   water.)
   Alternatively, in summer, circulate cold for a
   week, then turn heat on for 6 hours only.

Boiler — descaling

5 Drain and rinse three times as follows: tie-up ball-valve in header tank and drain the hot solution out with radiator vents closed, release ball-valve and re-fill with cold water, then re-open vents. Drain, then bail out any water left at the bottom of the tank.
6 Re-fill.

Note: When the system is clean, corrosion inhibitor can be put in to prevent future trouble.

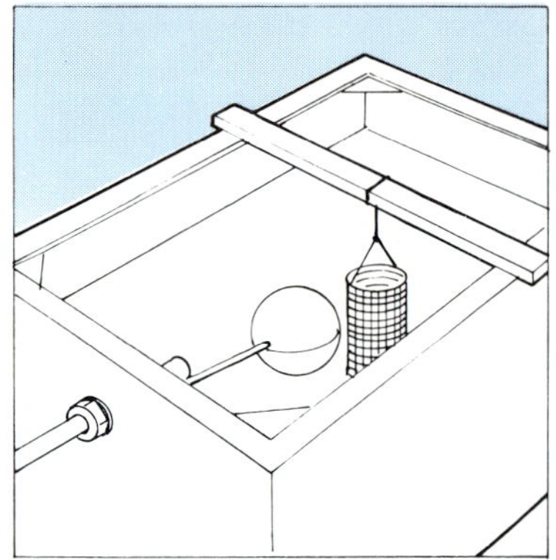

Waste-pipes

Symptom
Water unable to flow away, or flowing sluggishly

Cause
Blockage from solid matter and/or grease. Avoidable causes include: flushing disposable paper goods down WC; allowing hairs from shampooing to go into washbasin drain; pouring fat or cooking oil down sink drain; failing to add a sink-strainer to drain-hole or to use sink-tidy for food scraps. For frozen waste-pipes, see Waterpipes, page 32. For main drains, see page 72.

Tools and materials needed
Alternatives, depending on method chosen, include: washing soda, caustic soda, spirits of salts, flex or steel-tape drain-clearer; plunger; large spanner and bucket; hosepipe and clip or connector.
Some waste-hole grids can be removed with a screwdriver.

De-scaling units

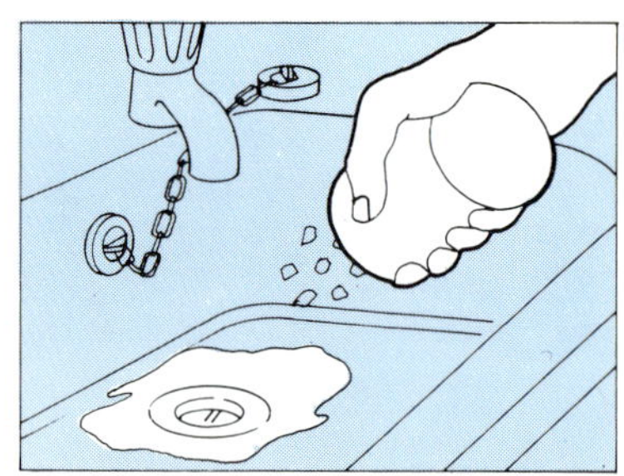

1 Soda and boiling water

Waste-pipes — blockage

Alternative methods

Boiling water and washing soda may clear minor blockages.

Caustic soda (sodium hydroxide) can be used in jelly form to dissolve material blocking WCs. First, bale out as much of the trapped water as you can. Empty the tin into a gallon of cold water, stir well, and pour slowly in. Leave for $\frac{1}{2}$ hour before flushing it away. Repeat if the first application was ineffective. Caustic soda crystals are less powerful but adequate for many sink and basin blockages (and to clean overflows made smelly by a build-up of soap–alternatively, use a wire-handled bottle brush). Having baled out what water you can, put 3 tablespoons of crystals in the wastepipe, followed by a cup of hot water. Flush away after $\frac{1}{2}$ hour. Be careful not to splash: caustics can damage textiles, paintwork or aluminium and will injure skin or eyes. If an accident occurs, flush the injury with plain or (better still) salt water. Any eye injury should be reported to a doctor instantly. Store the can out of children's reach. Other chemical solutions for this problem are 'Elsan' fluid, sold for caravan WCs; or 'Disclean', sold for cleaning masonry.

3 **A flex drain-clearer** is similar to curtain flex (from which a drain-clearer can in fact be improvised). It consists of about a metre of spiral wire with one end opened up like a corkscrew. This end is gently inserted into the blocked wastepipe or overflow of a basin or sink, while turning the handle at the other end in a clockwise direction. When the obstruction has been shifted, by pulling more than pushing, continue turning while withdrawing the wire. It may help to use the flex from the other end of the waste-pipe if this is accessible nearby outside the house.

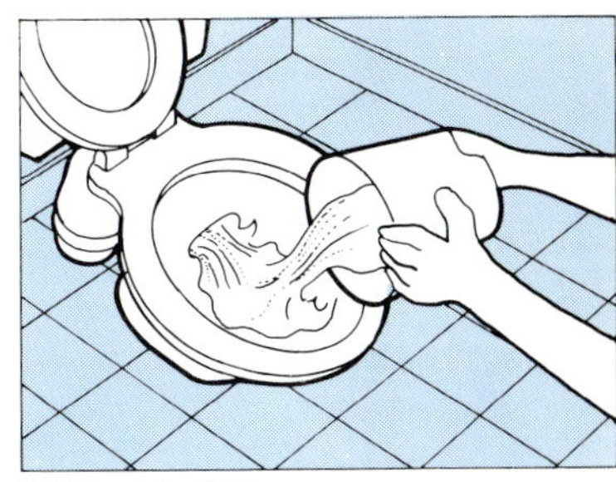

2 Caustic jelly

Waste-pipes — blockage

4 **A steel-tape drain-clearer,** also for use in sink or
basin waste-pipes, is much longer and can usually
penetrate right through to the outside end of the
waste-pipe. By pushing and then pulling, while at
the same time twisting the tape by means of a grip
provided at the other end, the conical front of the
tape can be worked past the blockage and then
used to free it. Finally, wind it back. The 'Sani-
Snake' is another flexible drain clearer, more
powerful; it is obtainable from hire shops.

5 **A plunger** consists of a rubber or flexible plastic
cup on a handle. Larger sizes have a disc above to
spread the pressure on the cup and to prevent it
from turning inside out. More expensive pump-
type plungers are also sold. A plunger can also be
improvised: tie a plastic bag round the head of a
mop, or round a sponge tied to a stick; or screw the
rubber disc from a power sander to a stick. When
using a plunger in a sink, bath or basin, first stop
up the overflow with a wet cloth or sticky-tape so
that air cannot escape this way. Remove all but a
little water from the basin, grease the rim of the
rubber cup, and place it over the waste hole. Pump
vigorously up and down. When the pool of water
disappears the blockage has been shifted.
In the case of a WC, a long-handled plunger is
needed and the action needs to be swift and
energetic, but not so vigorous as to crack the pan.

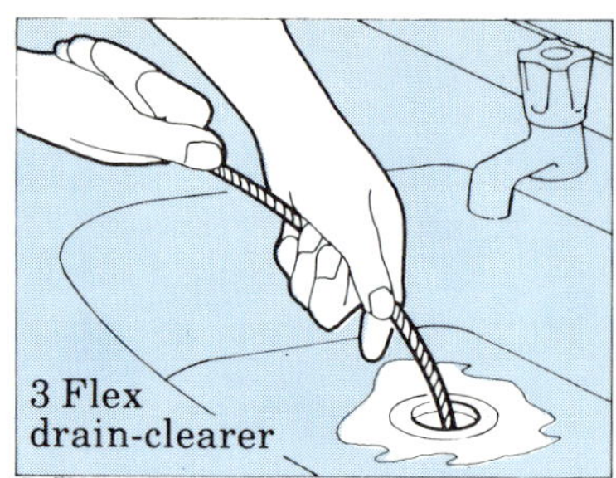
3 Flex drain-clearer

5 Plungers

Using a plunger

Clearing WC

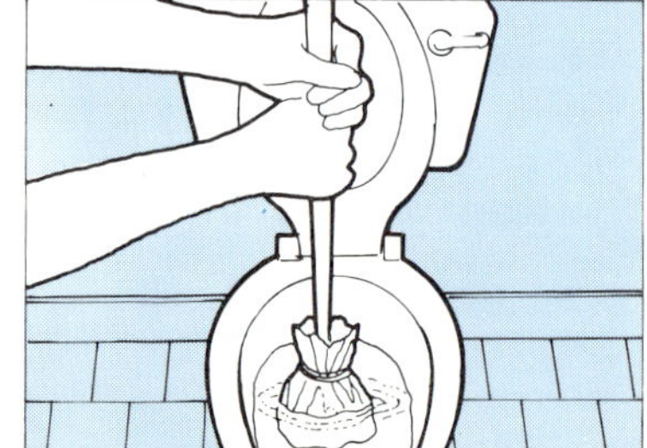

Waste-pipes — blockage

A large spanner may be needed if it proves necessary to open up the trap in the waste-pipe below the sink, basin or bath. (See advice about spanners and release-lubricants on page 17.) Before starting, place a bucket below the trap to catch water (or a tray in the case of a bath trap) and put the plug in the waste-hole.

U-traps

a Undo the screw-plug at the bottom of the trap using a screwdriver between the lugs, or else using a spanner. Hold the U steady with the other hand, otherwise you may pull it out of shape. In the case of a bath, you may need an angled bath wrench to reach it.

b If the obstruction does not then clear itself unaided, use a flex drain-clearer inserted from below, directing it into first one half then the other half of the U.

c When screwing the plug back, grease the screw-thread first and do not over-tighten. If its washer is worn, replace it.

Alternatively (where there is no plug at the bottom):

a Use a spanner, if necessary, to undo the two nuts that join the tops of the U to the pipes.

b Use flex to clear the U and the ends of the pipes.

c Screw back carefully – forcing can ruin the screwthread

Bottle-traps

Unscrew the bottom half by hand and poke a flex up into the waste-pipe in both directions.

A short piece of hose-pipe fastened to a tap can be used to force water into a sink, basin or bath waste-pipe in the hope of clearing the obstruction, but it will be necessary to plug the waste-hole and the overflow with wet cloths as previously described.

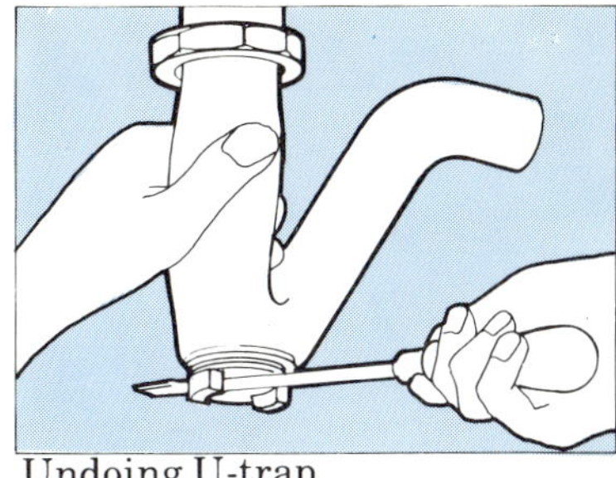
Undoing U-trap

Clearing U-trap

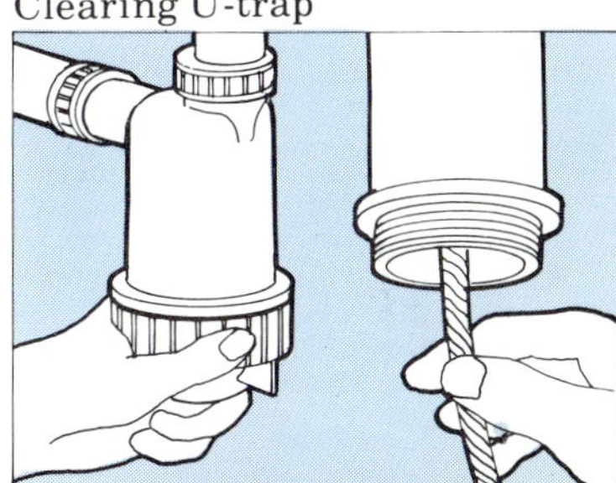
Undoing and clearing bottle trap

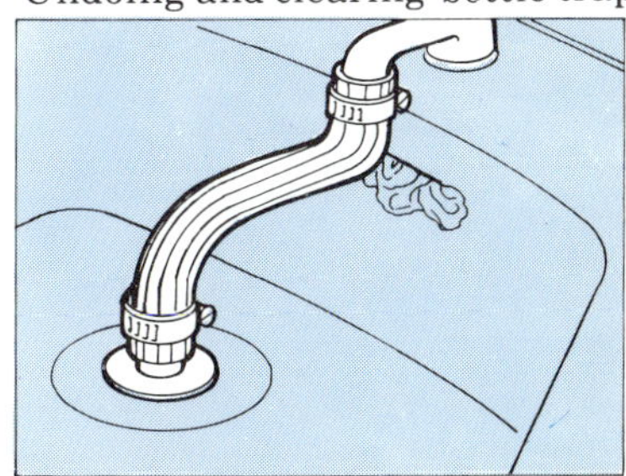
Hose-pipe method

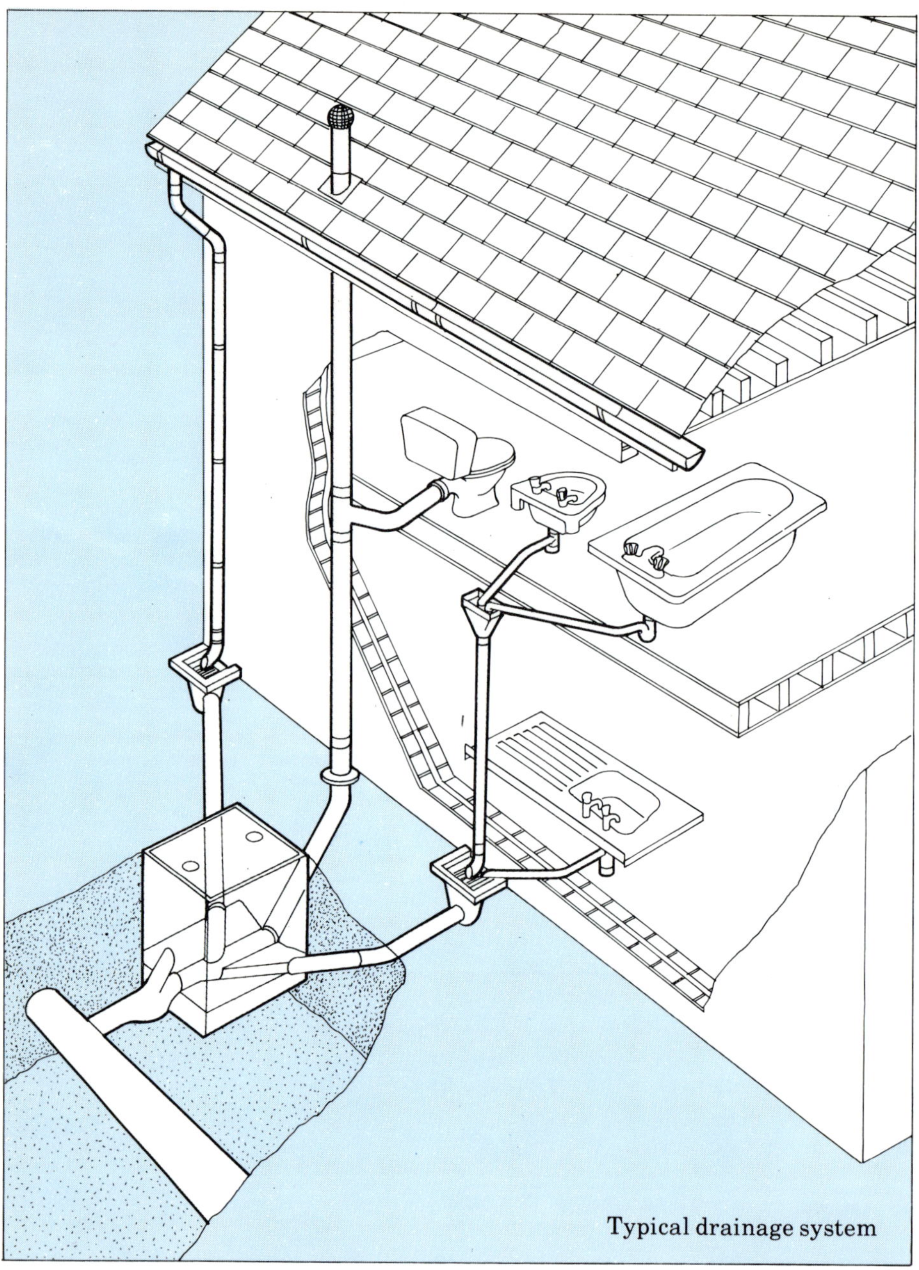

Typical drainage system

Waste-pipes — leaks

The 'Kynacolt' may help when all else fails. It has a cartridge of liquid gas which gives a burst of pressure. Can be hired.
Note
Do not leave cleaned waste-pipes unflushed, for it is necessary to keep their traps filled with water in order to seal off smells from the drains outside. Occasionally, siphonage keeps on emptying a trap: remedying this is a job for a plumber.
If the blockage is beyond the WC or wastepipe, see Drains, page 72.

Symptom
Leak

Cause
Crack
Or non-watertight joint

Tools and materials needed
Cracks: see Water pipes
Joints:
Large spanner,
New washer(s),
Vaseline
Or waterproof building tape and non-hardening sealant

Method
1 **Cracks** See water pipes.
2 **Joints** Disconnect (see page 16) and replace old washer(s) with new. Grease screw-threads before reconnecting.
3 **WC joint** Rake out old filling, bind tape round outlet from WC and thrust hand into socket, push sealant in, cover with more tape.

Gulleys and drains

Symptom
Overflowing gulleys

Cause
Blockage

Tools and materials needed
Trowel
Washing or caustic soda
Possibly a new grid
Rubber gloves

Method
1 If the blockage is in the gulley itself, remove and clean the grid, and dig any leaves or debris out. Use washing or caustic soda as described on page 67, probing with a stick to help loosen the blockage. Silt at the bottom is best removed by hand.
2 If the blockage is in the drainpipe below the gulley, use drain-clearing rods (see page 68).
3 If iron grid is broken, replace with new plastic one.

Symptom
Water in drains unable to flow away, or flowing sluggishly. Inspection pit in garden overflowing.

Cause
Blockage

Tools and materials needed
Drain-clearing rods and attachments (from hire shops)

Method
1 Remove cover from pit nearest house. If it has water in it, blockage is further on. Remove cover from next pit if there is one; if this has water, blockage is still further.

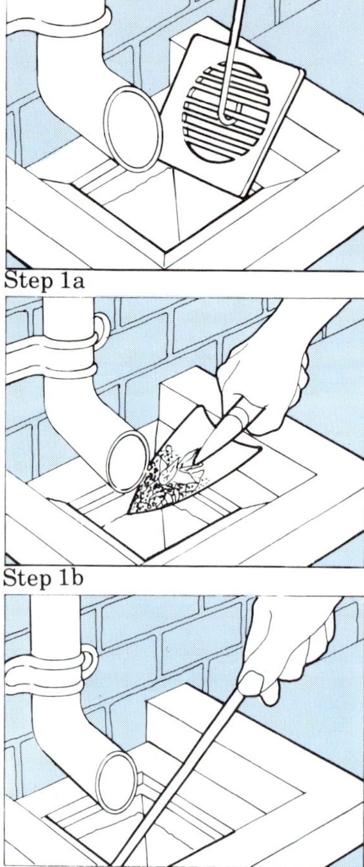

Gulleys and drains

If pits are dry, blockage is within one of the waste pipes coming from the house to the first pit. To identify which one, send water down each in turn and observe which lot fails to reach the pit. (Some houses have only one waste-pipe.)

2 Working from an empty pit, screw the flexible rods together one at a time while pushing them up the pipe that has the blockage. Turn clockwise only, otherwise you will unscrew the rods. The first rod should have the corkscrew attachment on it to loosen the blockage; the rubber plunger is attached instead when pulling is needed.

3 If you have to work from a full pit, bale this out to locate the outgoing pipe (centre bottom of the side nearest the road) and try to clear this with the plunger. Failing that you will have to take the stopper out of the 'rodding eye' (a small opening above the outgoing pipe) and rod through this. Sometimes it is this stopper that has fallen and is causing the blockage.

Note

Specialized drain-clearing firms using powered rods or pressure jets can be located through the Yellow Pages: ask for a quotation before going ahead, and a guarantee. Alternatively, ask the local Environmental Health department of the council whether they will clear the drain or recommend a firm. A few councils provide this service free. Jet-operated drain clearers can be hired.

Where a drain serves more than one home its clearance will, if it was laid before October 1937, be treated as a public responsibility though the council may bill the householders concerned for doing it. Shared drains laid since then are the responsibility of the householders concerned. If you call in a drain-clearing firm without prior agreement with others, you will be liable to pay the bill.

Drain rods and attachments

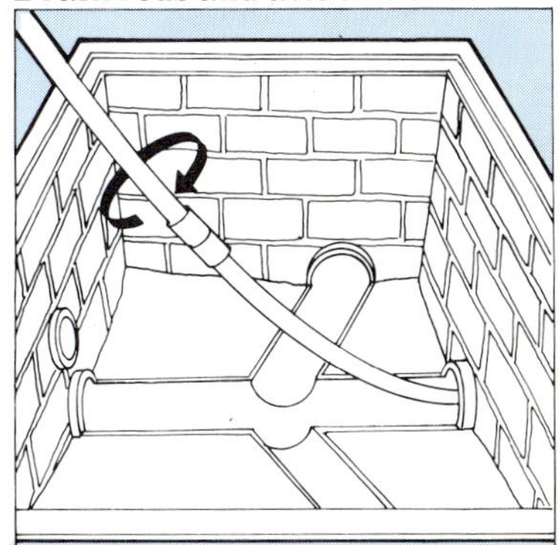

Step 2

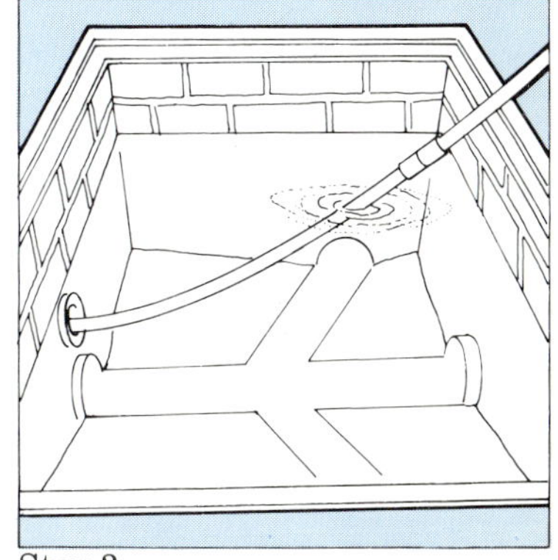

Step 3

Gutters and rainwater pipes

Rust prevention

Iron gutters and pipes need regular painting. Use bituminous paint inside gutters, such as Aquaseal 44 or Presomet. Jenolite (see Water tanks, page 41) both removes and inhibits rust, provided it is painted over, so does Kurust.

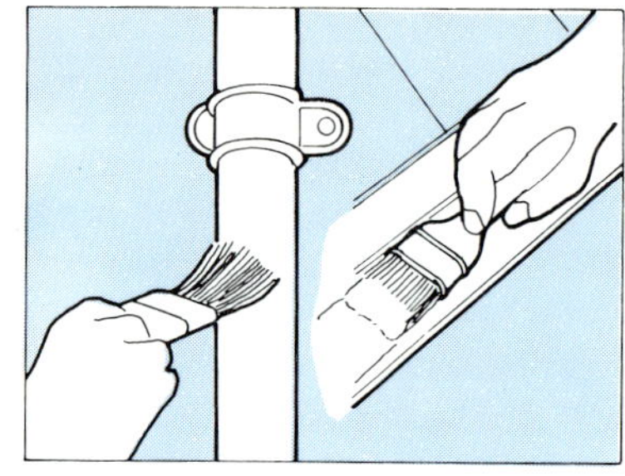

Symptom

Overflows and leaks, stains on wall

Cause

Blockage in gutter; cracks or holes due to rust

Tools and materials needed

Ladder
Stiff brush
Long stick
For cracks: waterproof building tape (eg Sylglas or Aquaseal Flashing) and scissors
Or waterproof filler

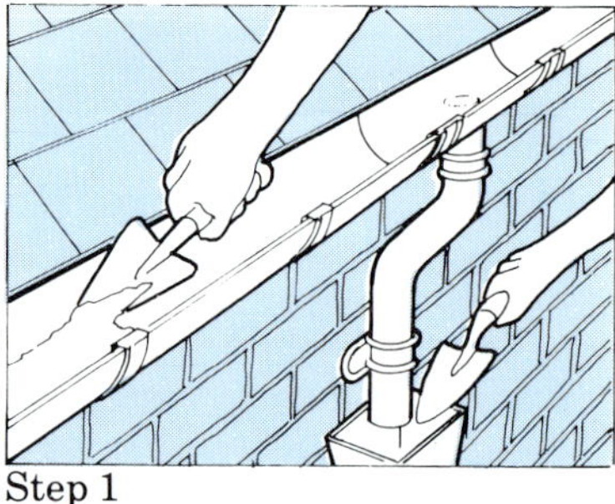

Step 1

Method

Choose a dry day, without frost, for this work.
1 Brush leaves and debris out of gutters.
2 If possible, remove any angled pieces joining gutters to downpipes and clear these out; clear any hopper heads.
3 Thrust a long stick (with rag tied to its end), or a hose, down pipes. Have a pan ready at the bottom to catch debris).
4 Many cracks can be mended by binding waterproof tape round the pipe or along a gutter (buy an adequate width for the cracks, keep the tape in a fairly warm place, and do not stretch it as you apply it). Clean the surface well first, and overlap the strips of tape (which should extend beyond the actual crack.) Or use a waterproof filler (see page 32) or else a non-hardening mastic such as Synthaflex or Aquaseal 88.

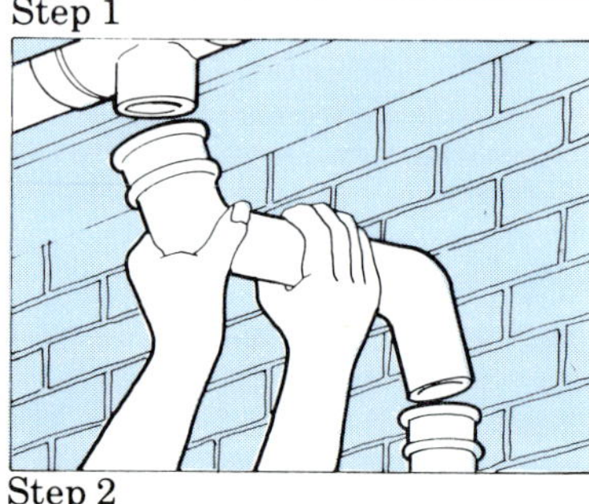

Step 2

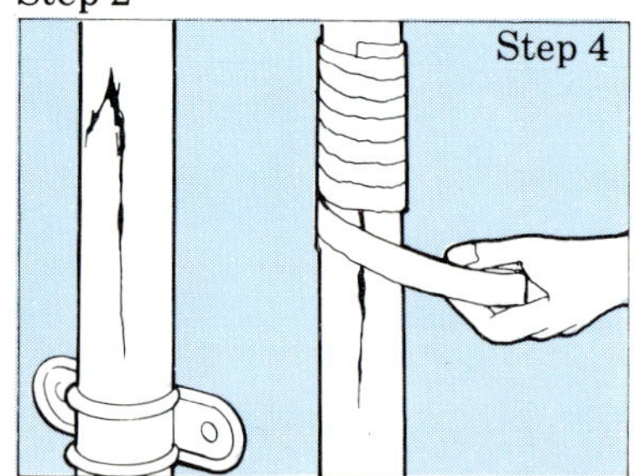

Step 4

Gutters and rainwater pipes

5 If a small part of a gutter has broken off, tie cardboard over the gap, grease the inside and coat thickly with waterproof epoxy filler such as Isopon (let it overlap the gutter by 50mm or more). When it is hard, remove the cardboard.

6 If mesh guards are missing from tops of downpipes, these should be put on. Gutters can be kept free of leaves by clipping on a special gutter grid made of rotproof mesh.

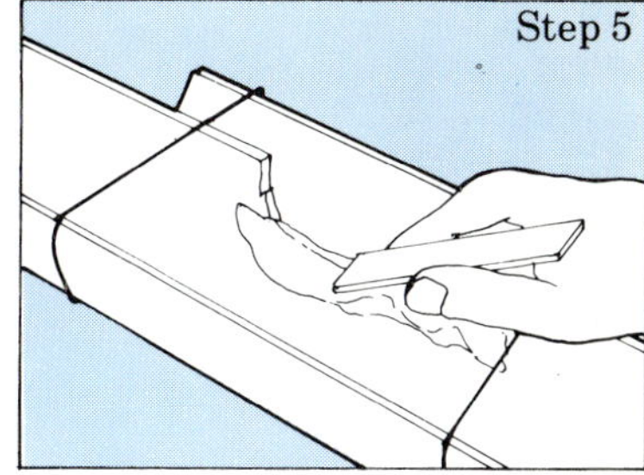

Note: There may be gutters not visible from below: between two pitched roofs, or around flat roofs.

Gutters and rainwater pipes

Symptom
Gutters leaking at joints

Cause
Deteriorated sealing material

Tools and materials needed
Ladder
For plastic gutters:
Petrol
New rubber seal
For iron gutters:
Spanner
Screwdriver
Putty
Putty knife
Possibly, hacksaw, bolt, washer and nut

Method
For plastic gutters
(Long sections are usually connected by a short
one, called a union, which is lined with sponge
rubber seal.)
1 To replace the seal, unclip the union and take it to
  the retailer to get a new seal of the right size.
2 Use petrol or lighter fluid to clear off all remains
  of the old seal.
3 Insert the new one and squeeze the gutter ends
  in order to slip the union back on again.
For iron gutters
(Long sections usually overlap one another and are
fastened with a bolt, sealed with putty.)
1 Using a spanner and screwdriver (and possibly
  dismantling lubricant), remove the bolt.
2 If it is immovable, saw the nut off.
3 Prize the two sections apart and scrape out old
  putty.
4 Spread new putty in, press the two sections
  together.
5 Secure with bolt, washer and nut.
  Alternatively, treat as for cracks, see page 20.

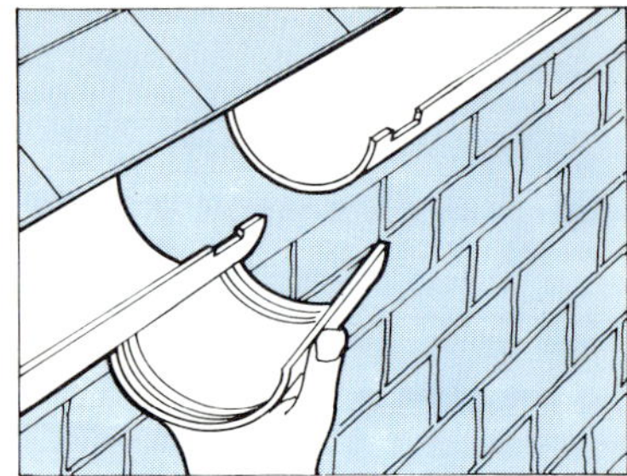

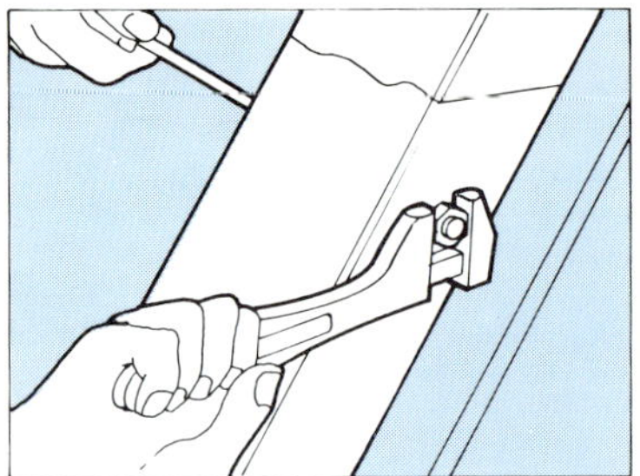

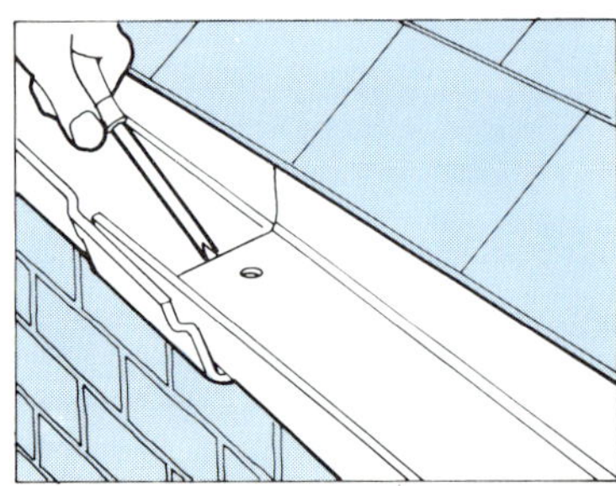

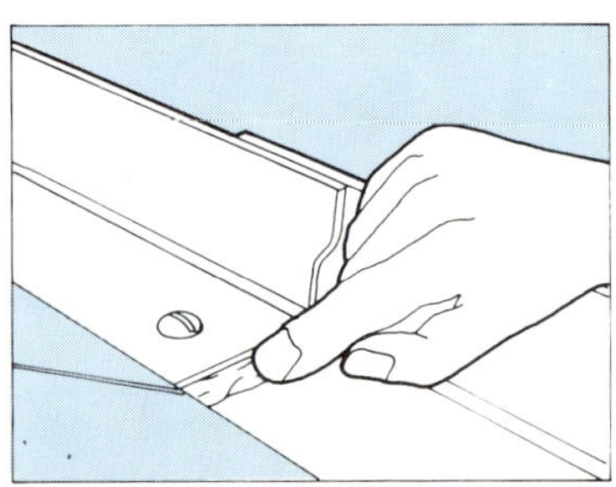

Gutters and rainwater pipes

Symptom
Gutters dripping from middle

Cause
Sagging

Tools and materials needed
Ladder
Bradawl
Screwdriver
Hammer
Large nails

Method
1 Clear out debris that has accumulated in the sag.
2 Lift the sagging section to the correct level again. The gutter should slope slightly but steadily down to the downpipe at one end.
3 It may be necessary to refasten its brackets again in a new position: no problem if they are simply screwed to the fascia board behind the gutter but many are secured to the ends of rafters, inaccessible behind the tiles. Very long nails hammered into the fascia board may give sufficient extra support to prevent the sag recurring, or wedges between gutter and brackets.

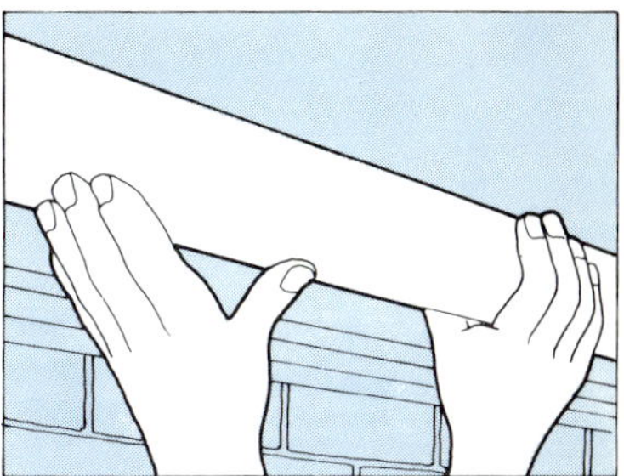

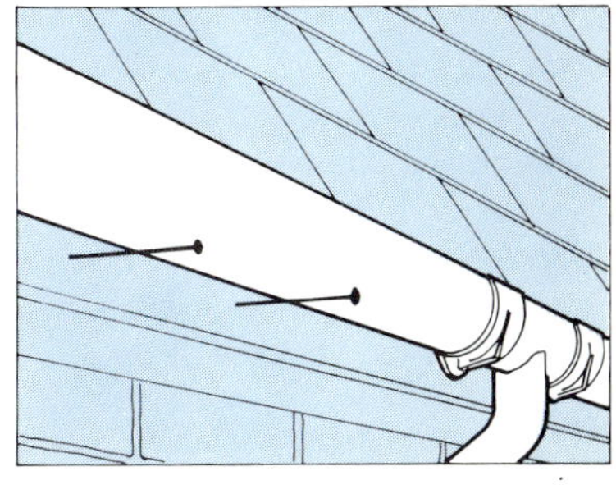

Note: If the sag is more than about an inch, rectifying it may break the seal between sections, which will then need renewing as described on page 76.

Gutters and rainwater pipes

Symptom
Rain-water pipes leaking from cracks

Cause
Pipe not held securely; or pressure of debris that
has entered through unsealed joints

Tools and materials
Ladder
Claw hammer
Punch
Epoxy putty
Waterproof mastic filler

Method
1 To secure a wall bracket that has loosened, remove
each section of pipe below the bracket. This is done
by unfastening the brackets holding them, using a
claw hammer and piece of wood to lever their nails
out as shown. Repair any cracks as on page 20, or
buy new section if necessary.
2 To fasten the wall brackets anew, use a hammer
and punch to make large holes in the mortar, fill
with wedges of wood and hammer in the nails
through the brackets.
3 Clean any debris out of the pipes before replacing
each section, fill the joints with mastic.
Do not seal the joints hard — you may want to
disconnect them in the future.

Replacement
Because of their light weight and easily clipped-on
fastenings, plastic rainwater goods are easy to
install. Lifting down the heavy old iron ones is the
effortful part of the job.

75mm gutter is usual for sheds, porches and
garages; 100mm for houses. Pipes are 50 or 65mm.
Only the larger ones are available in black or
white as well as grey.

Assemble all the parts on the ground first. Then

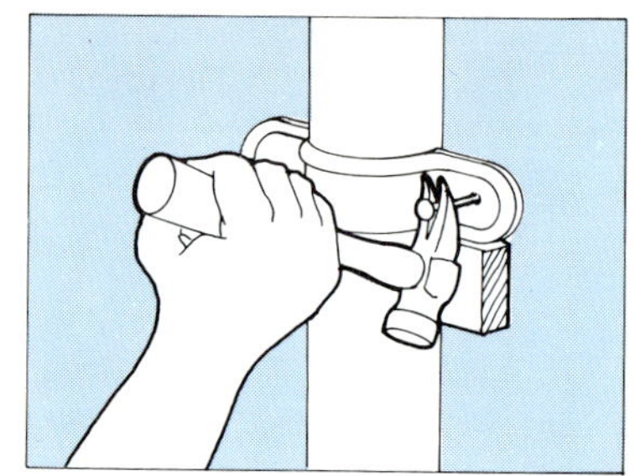

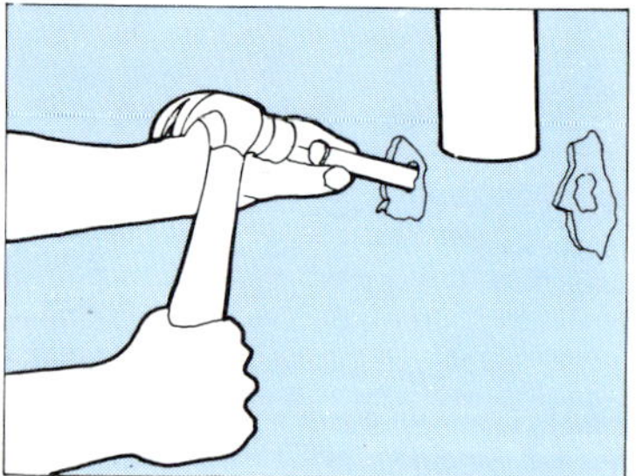

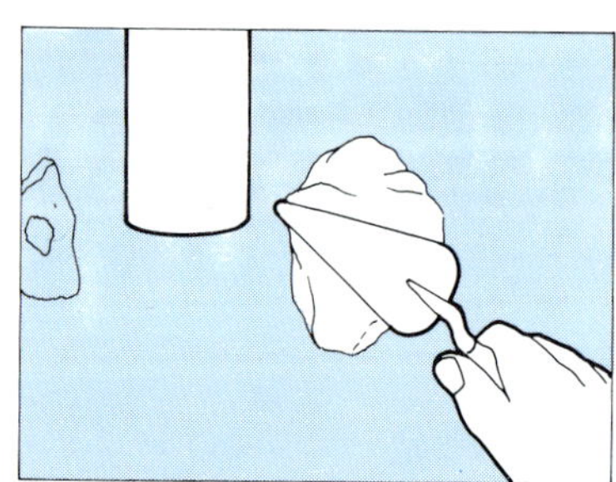

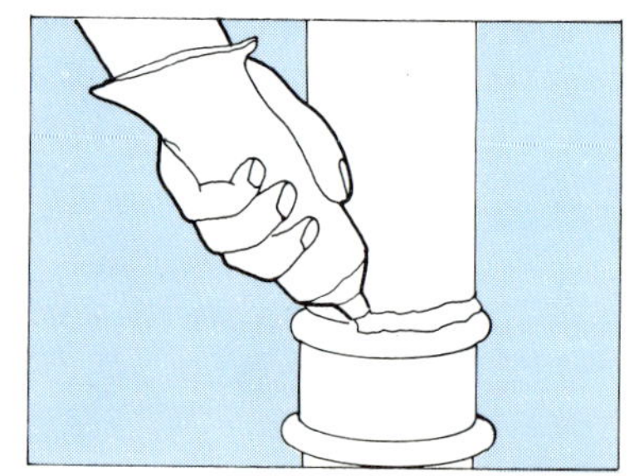

Gutters and rainwater pipes

mark where the gutter brackets are to go on the
fascia board (which runs along the top of the wall)
and screw these on. They should be not more than
1 metre apart; leave no join in the gutter
unsupported by a nearby bracket. Some houses
have no fascia boards, so special brackets are
needed to screw to the ends of rafters under the
tiles. The gutter should slope slightly towards its
outlet, so nail on a piece of string as a guide before
fastening the brackets.

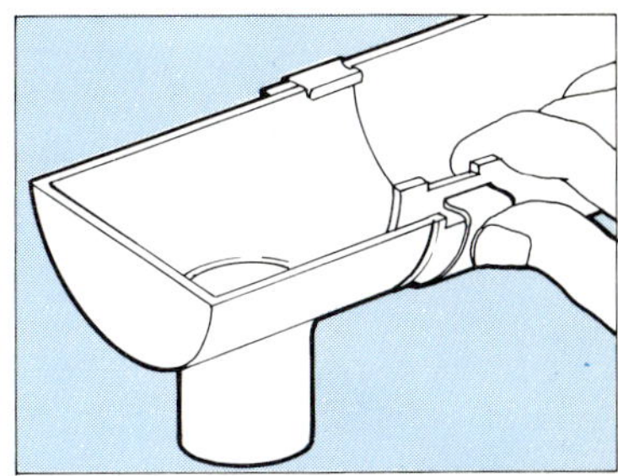

To attach the end outlet to a length of gutter
Clip the notched end of the gutter inside the outlet,
secured by a gutter strap.

To join two lengths of gutter Clip the end of one
inside the other, secured by a gutter strap.

To shorten a length of gutter Use a fine-tooth
saw to cut, and a file to make new notches at the
sawn end if necessary.

To close the end of a gutter Clip on a stop,
secured by a gutter-strap.

To attach a pipe to a gutter outlet An angled
pipe, called an offset, is usually needed, pushed
onto the end of the outlet. The pipe is then pushed
into the offset. If more than one length of pipe is
needed to reach the ground, the lengths are simply
pushed together.

To secure a pipe to wall Clips encircling the
pipe are screwed to plates behind it and these are
nailed to the wall. Usually each 3-metre pipe needs
3 clips.

To terminate the pipe Unless connected to an
underground drain, the pipe needs a 'shoe' at the
bottom, directing the rainwater into a gulley or
water butt.

In addition to these components, there are various
clip-on bends and angles for special situations.
Downpipe adaptors and waterbutts can be used to
conserve rainwater.

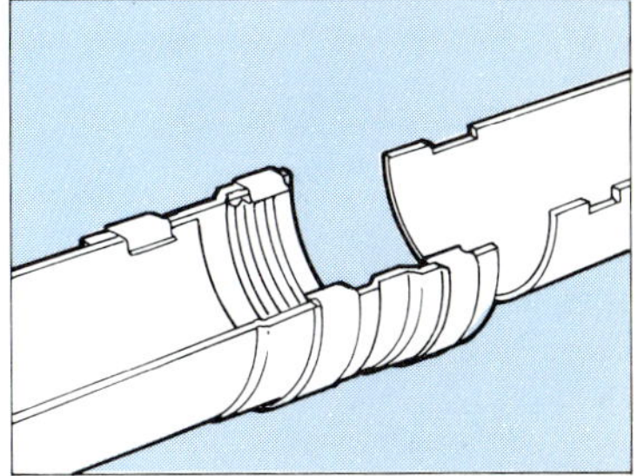

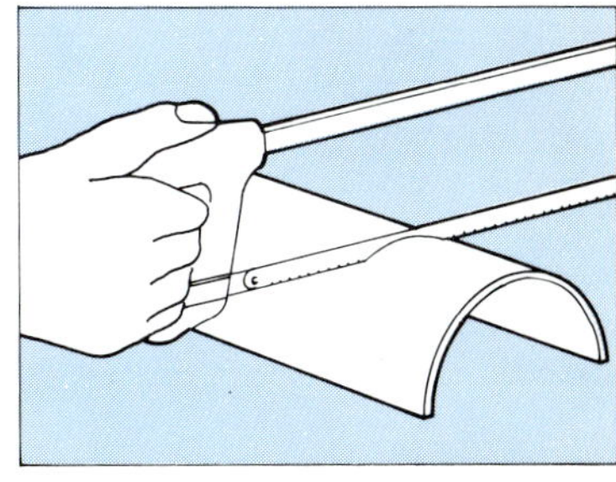

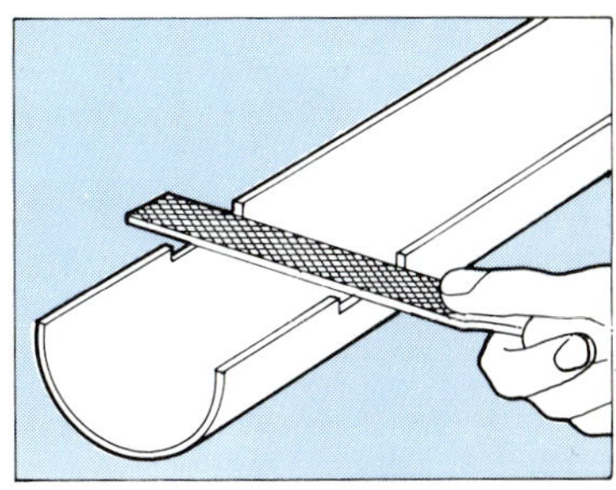

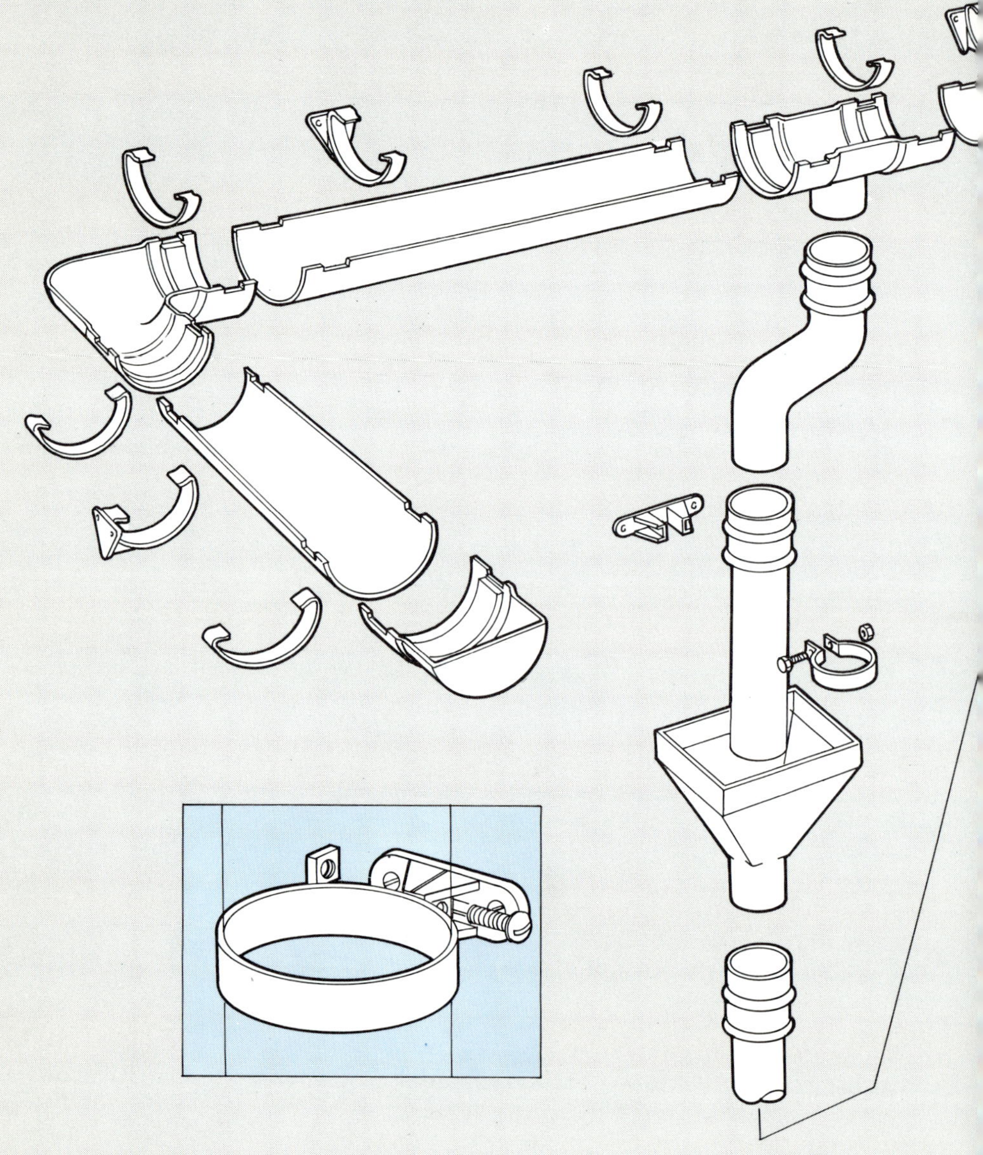

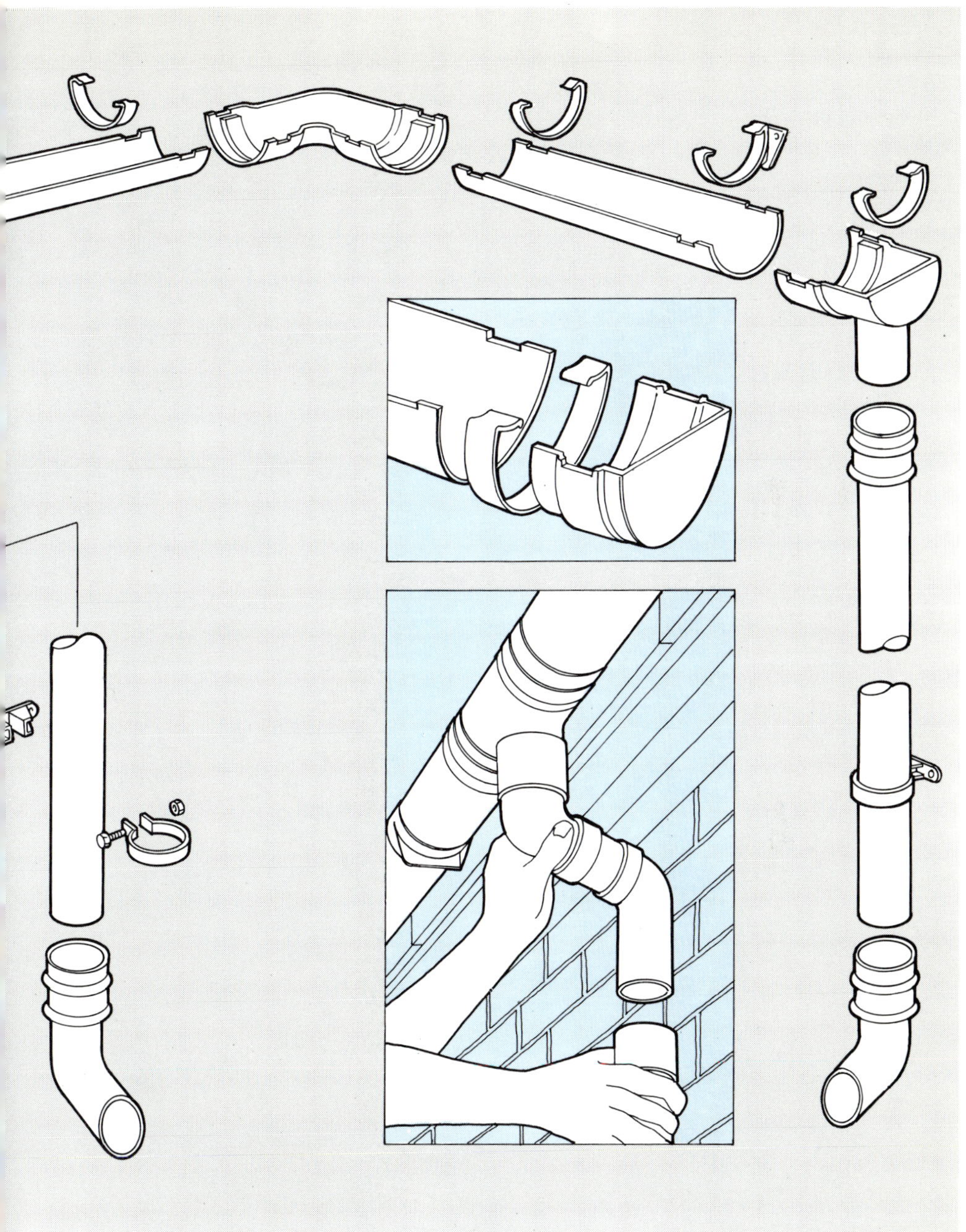

Ladders

(These may be hired: ask at several shops because charges vary a lot).

To get at gutters you will probably need an extendable ladder, rope-operated if it is very long. If the ground is not level, wedge one of its feet. Tie the bottom to something nearby to prevent any slipping. The diagrams show how to raise a ladder, the best angle (75°) at which to place it, and how to use a special stay if the roof has a large overhang (do not rest a ladder on plastic or frail old gutters). There are accessories to stop feet slipping and to provide a platform to stand on, which may be hired too.

Aluminium ladders are often lighter than wood, very robust and have non-slip rungs. For comfort, rungs should be flattish and wide. Never use a damaged ladder.

Because thousands of accidents occur with ladders every year, and dozens of deaths, it is worth observing safety rules:

Be sure you use the ladder the right way up, the right way round, dead straight, and with hooks fully engaged before you get on it.

Don't put it in a danger-spot (on sloping or slippery ground, in front of a doorway, in a windy spot).

Keep one hand always holding a rung. Use a bucket for tools with a hook to hang it from a rung.

Keep your body in line with the ladder – don't lean sideways, nor put one foot on a windowsill.

Levelling and securing

Raising a ladder

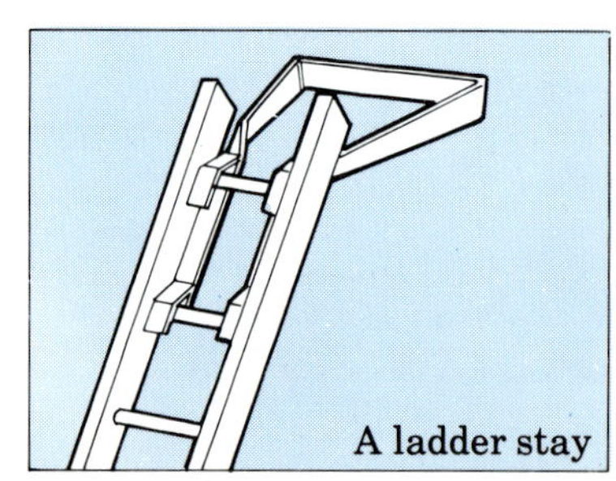

A ladder stay

Improvements

This book is primarily about repairs and replacements but there are now a number of improvements an amateur can install. Putting in new fitments has been made much easier recently with the advent of simple pipe–connections needing no blow–lamp, plastic joints needing no sealant, and pliable pipes which can be bent by hand. The problems of detaching old equipment were dealt with on page 78.

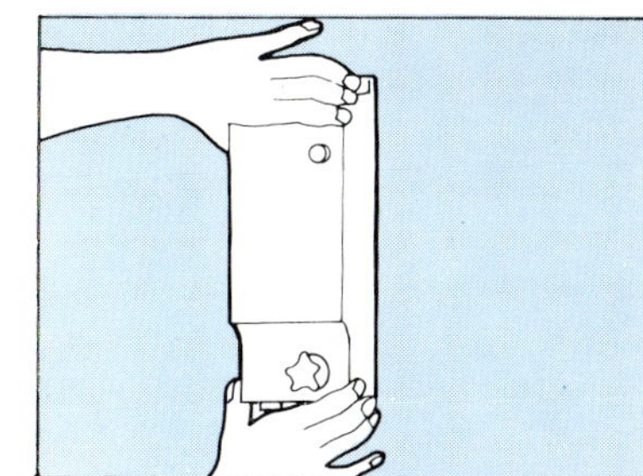

Installing a shower
(An electrician will be needed for the wiring.)

Tools and materials needed
Shower kit, such as Deltaflow
Drill
Wallplugs
Screws
Hacksaw
File
Spanner
Copper pipe (15mm)
T–joint, compression type
Possible, pipe clips

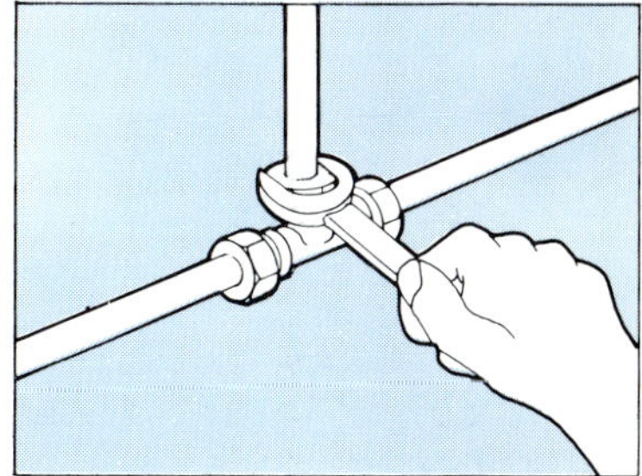

Method
1 Drain cold water supply (see page 12).
2 Decide on the position for the shower–head (eg on the wall beside the bath). The heater can go anywhere provided its top is on a level with the shower–head and that it is out of range of the spray of water.
3 Drill holes and screw the heater (with its cover off) to the wall.
4 Saw through the nearest length of cold water–pipe and file its ends smooth before inserting the T–joint, tightening this with a spanner. Fasten the new length of pipe to the T and, unless the pipe–run is short, secure it to the wall with the pipe–clips.

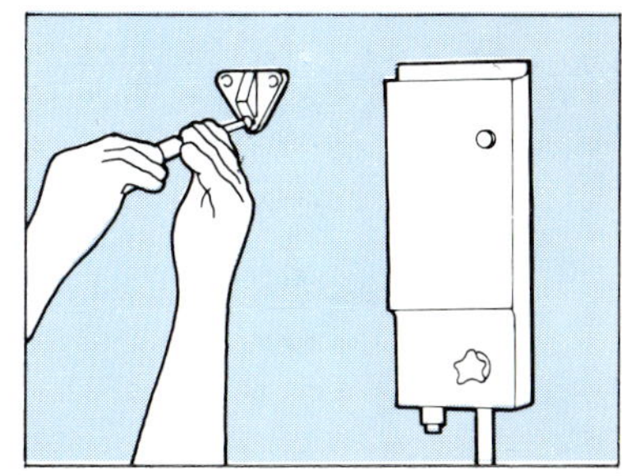

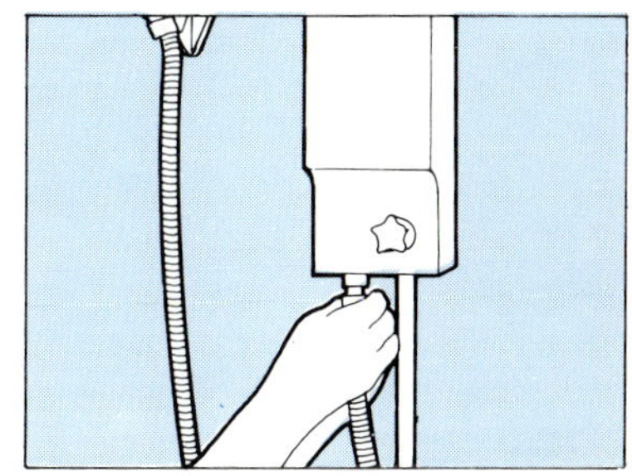

Improvements

5 Where it is not possible to run a new length of pipe straight up to the heater, several lengths and more joints would ordinarily be needed but it is easier to use Copperbend pipe which can be bent by hand to turn corners; the 15mm version of this comes in 400mm lengths, with a connector at each end.

6 Connect the other end of the pipe to the inlet of the valve inside the heater, and restore the water supply. Put the cover of the heater on.

7 Screw up the bracket for the shower–head and put this in place, with its flexible hose connected to the heater outlet.

Installing a sink or basin water–heater

The method is similar to that for the shower. Small electric heaters contain a 3KW element and hold 7 litres ($1\frac{1}{2}$ gallons) of water. When this is drawn off, reheating a further 7 litres ($1\frac{1}{2}$ gallons) takes about 10 minutes. Small gas water heaters, giving a continuous but limited flow, are similarly plumbed–in, leaving the gas connection to be done by a gas engineer.

Improvements

Installing a WC flush-panel

An ordinary cistern can be replaced with a slim panel.

Tools and materials needed

Flush panel and matching flexible flush–pipe (various colours)
Screwdriver
Screws
Wallplugs
Spanner or wrench

Method

1 Drain cold water system (see page 12), then remove the old cistern and its flush pipe. Unscrew the water supply pipe.
2 Attach connector to new flush–pipe.
3 Moisten the other end of the pipe and push it into the outlet of the panel, through the O–ring provided.
4 Attach the pipe to the WC by means of the connector.
5 Screw the flush panel to the wall.
6 Re–connect the water supply pipe and restore the supply.

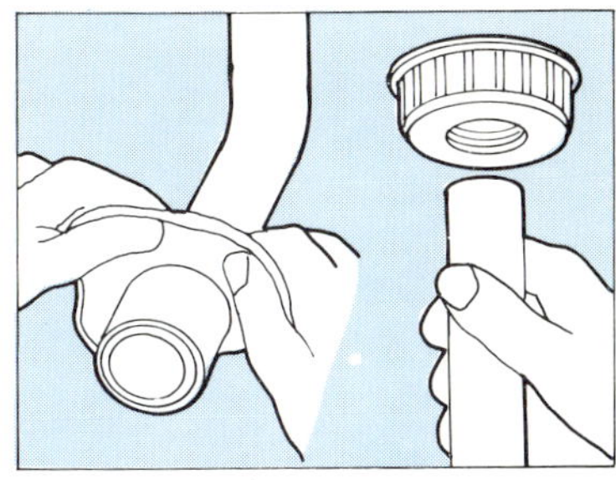

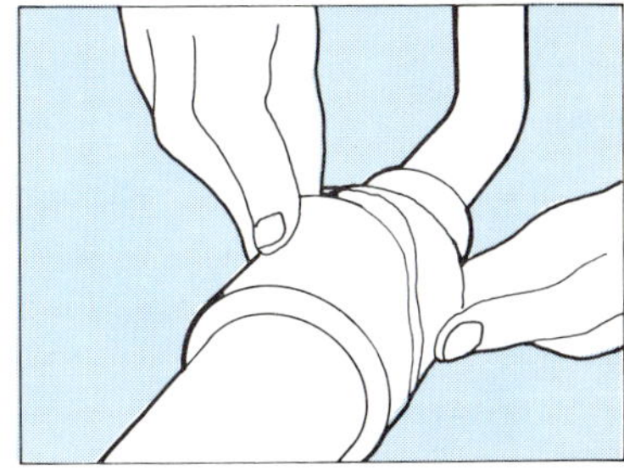

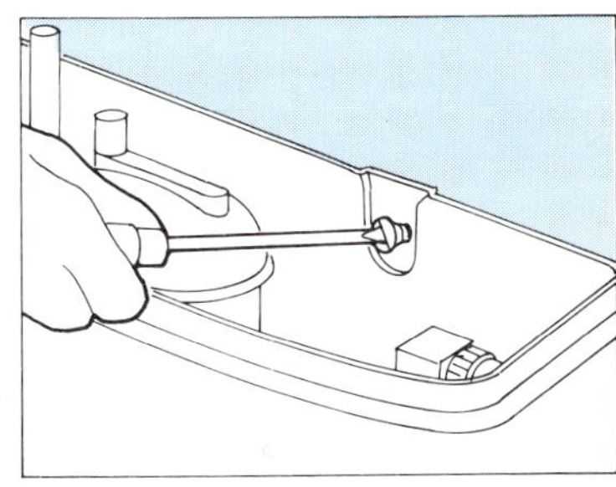

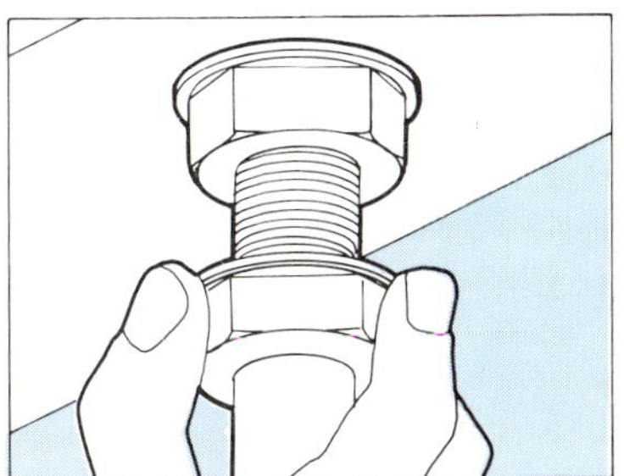

Improvements

Baths

Any bath with a square–cornered top can be given a more distinctive appearance by boxing it in with acrylic panels of curving shape, in a choice of colours. These 'Easifit' panels are adjustable in height and width so that almost any bath can be fitted. From the same manufacturer come press in plastic strips to seal the gap between bath and walls, and also a matching bath 'bar': a wall–hung rack for soap, bottles and so forth, to run the whole length of the bath. Thus an all–round transformation can be made with a screwdriver— without the expense and upheaval of a new bath.

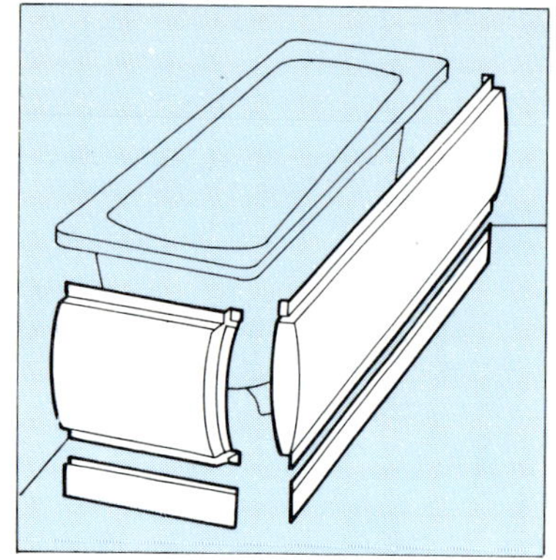

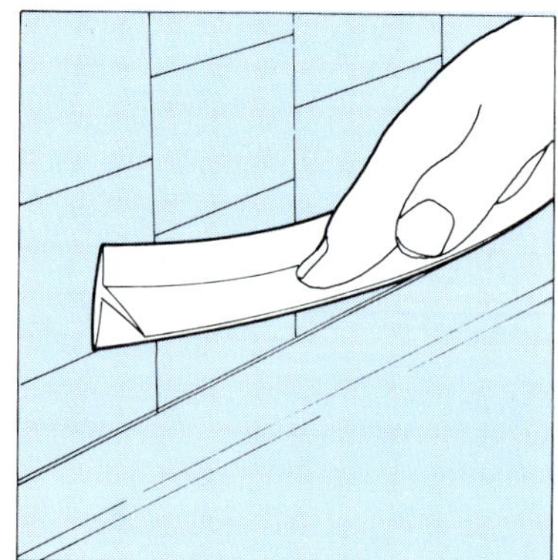

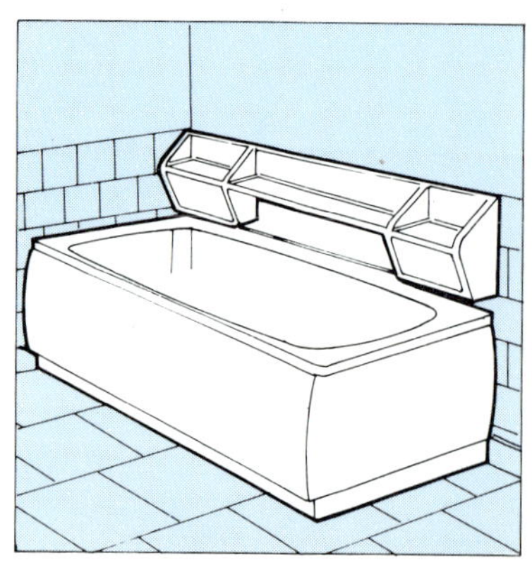

Improvements

Installing a mixer tap

Tools and materials needed
Basin wrench
Mixer tap (this will fit any standard sink with tap holes)

Method
1 Drain hot and cold water supplies (see page 12).
2 Unscrew the nuts connecting the two taps to the water pipes below the sink, and then unscrew the nuts holding the taps to the sink.
3 Put the mixer tap on in place of the two taps, pressing its gasket well onto the sink top.
4 Underneath the sink, fit the mixer's washers and nuts to its two tails, and tighten them.
5 Re–connect the water pipes to the tails and restore the water supply. If there are any leaks, tighten these nuts still more.

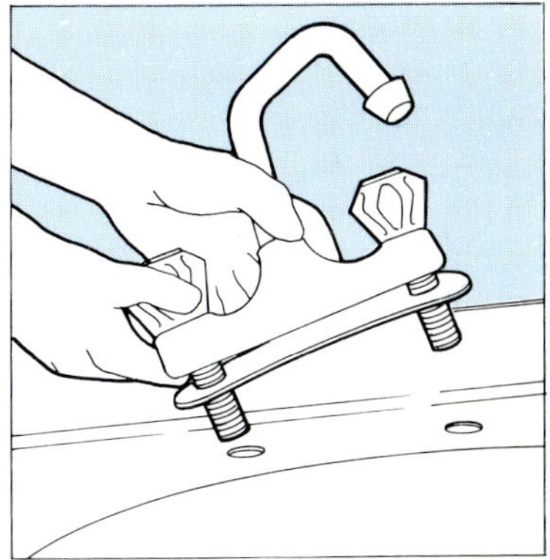

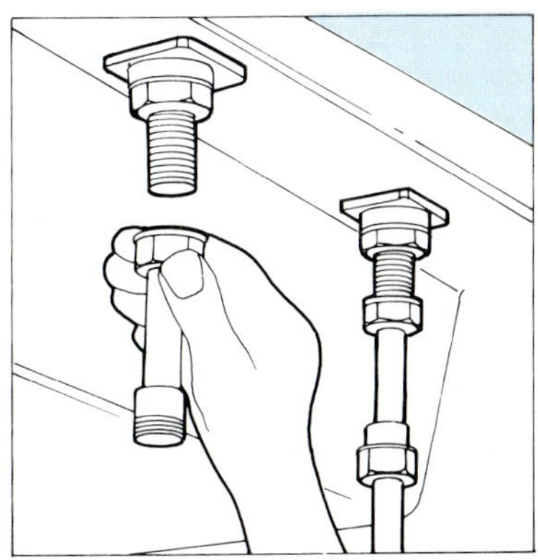

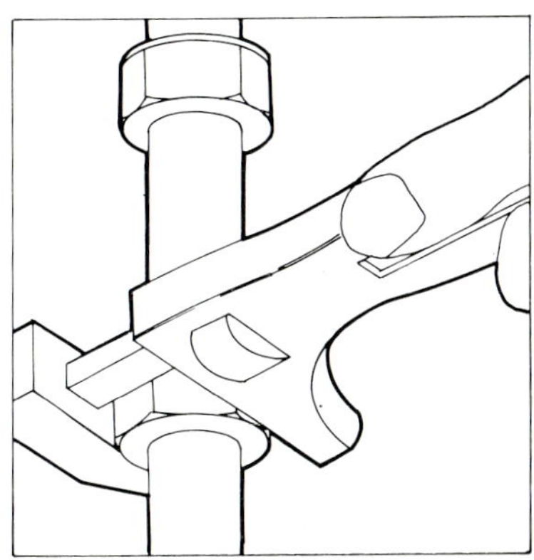

Improvements

Installing a sink spray

The purpose of this is to provide a combined brush and water–spray for cleaning dishes. An on–off lever on the head of the spray controls the flow.

Tools and materials needed

Hand drill
Screwdriver
Spanner
Gimlet
Spray kit

Method

1 Using a hole–cutter provided with the spray, drill a 22mm ($^7/_8$in) hole at the back of the sink top. (Unnecessary if the water pipe happens to run above the sink.)
2 Above the hole, place a washer and the holder for the spray–head, and thread the hose down through these.
3 Below the hole, another washer and nut are screwed into place.
4 In the hot water pipe below the sink, drill an 8mm hole on one side of the pipe only, after draining the water supply (see page 12).
5 Screw the two halves of the clamp together, over the hole in the pipe. (The clamp can then be screwed to the wall, if wished.)
6 Screw the end of the hose into the clamp and restore the water supply.

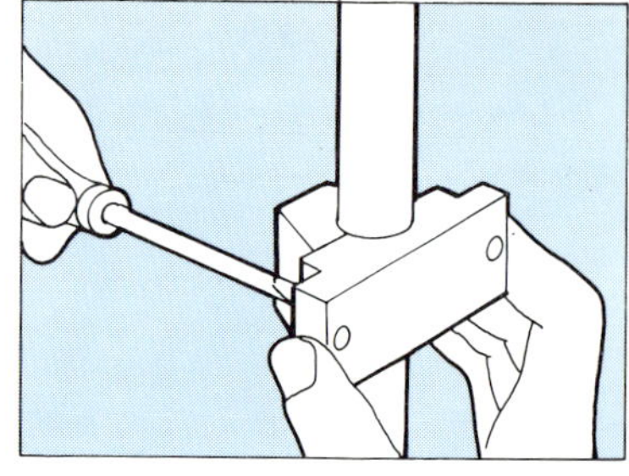

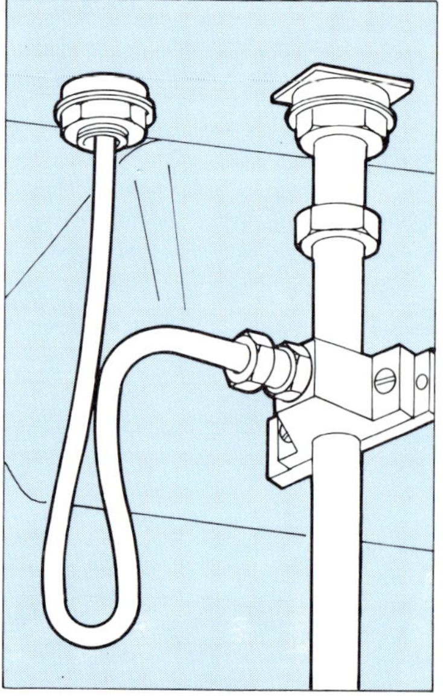

Improvements

Installing a garden tap

Tools and materials needed
Hacksaw
File
Spanner
Short length of 15mm copper pipe
Probably stopcock
Length of Copperbend pipe with tap connector at one end
Wallclips
Wallplugs
Screws
Screwdriver
Garden tap with nozzle for push–on attaching of hose
Long masonry drill (this can be hired)
Filler

Method
1 Drain cold water supply (see page 12.)
2 Insert T-joint, see page 91, into the cold water pipe under the sink or wherever is most convenient.
3 Connect a short length of pipe to the T–joint, then connect a stopcock to this (needed during frosty weather in order to isolate and drain the outdoor pipe; if you prefer to rely on lagging, see page 35.)
4 Drill a hole through the wall and pass a length of Copperbend through—one end connected to the stopcock and the other to the garden tap. Secure it with wall clips screwed to the wall.
5 Use an appropriate filler to close up the hole in the wall and restore the water supply.

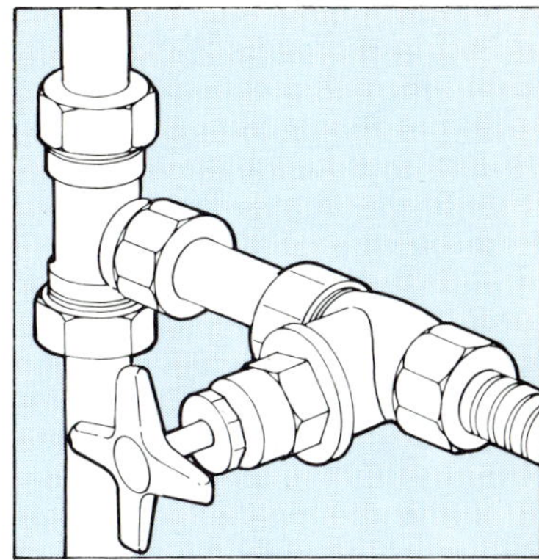

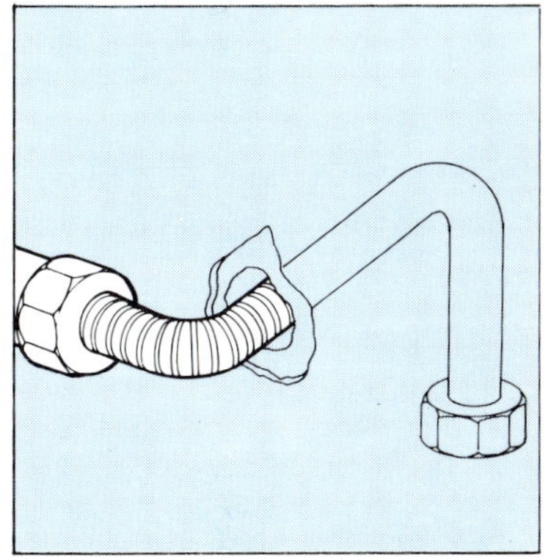

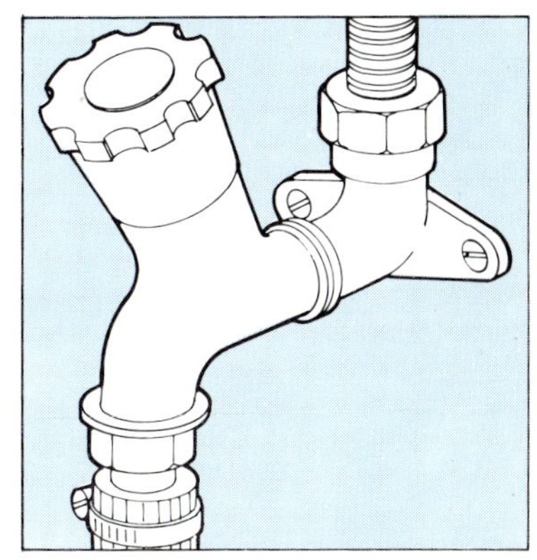

Improvements

Altering or adding to pipes

Many improvements involve work on pipes. In modern houses, only 15mm or 22mm (outside diameter) copper pipes are likely to be found—easy to deal with. If yours are of a different size, consult the retailer about joints for connecting new pipes to old. Where there are still old lead pipes around, a special adaptor is needed before any copper pipes can be joined on, and installing this is a job for a plumber.

Cutting copper pipes to length is done with a hacksaw and file (it may help if you have a vice to hold the pipe steady), then a compression joint with two nuts is put onto it in order to join one pipe to another. Joints may be of metal or plastic.

Pipes of plastic are for cold water only, and are far more vulnerable to damage. They may be rigid (join lengths with special adhesive) or flexible (use ordinary joints plus a liner). Flexible ones cost less.

Place the joint between the ends of the two pipes and hand-twist first one and then the other of its nuts to secure the pipe ends. Tighten with a turn of the spanner. If joint-sealing is needed, see page 71.

A good many angles and joins will be avoided by using pliable copper pipe (see page 83), which can be obtained already fitted with either pipe- or tap-connectors as required.

1 Pipe-connector: to make a straight join
2 Right-angle pipe-connector: to join one pipe at right-angles to another
3 T-joint: to take a branch pipe from the middle of a main pipe
4 Right-angle tap-connector: to join a tap at right-angles to the end of a pipe

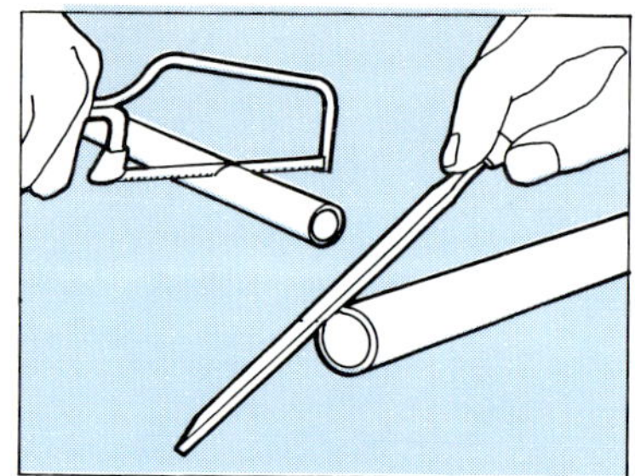

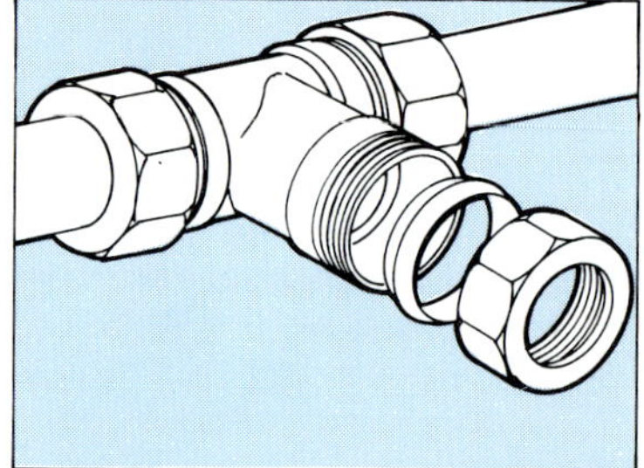

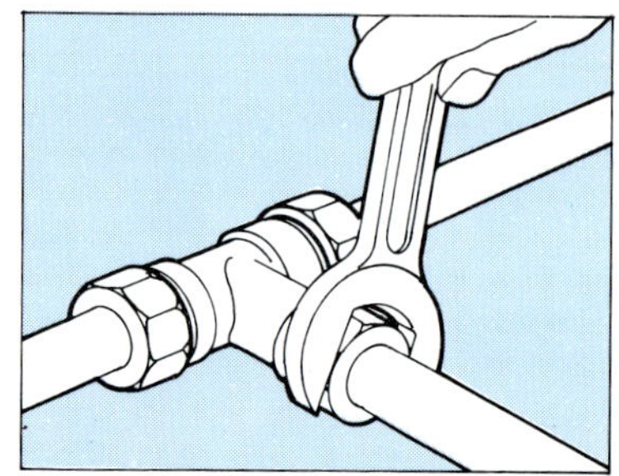

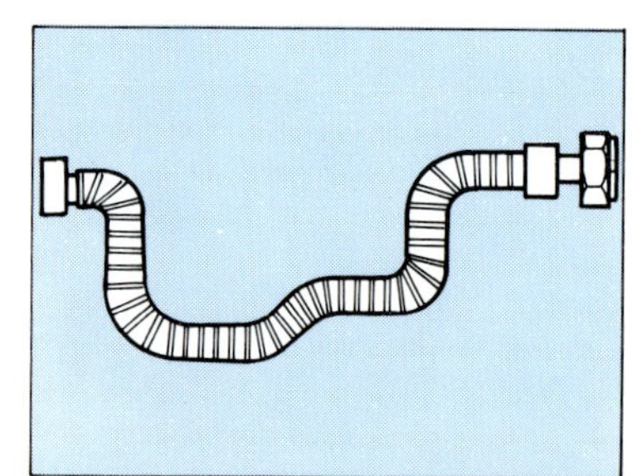

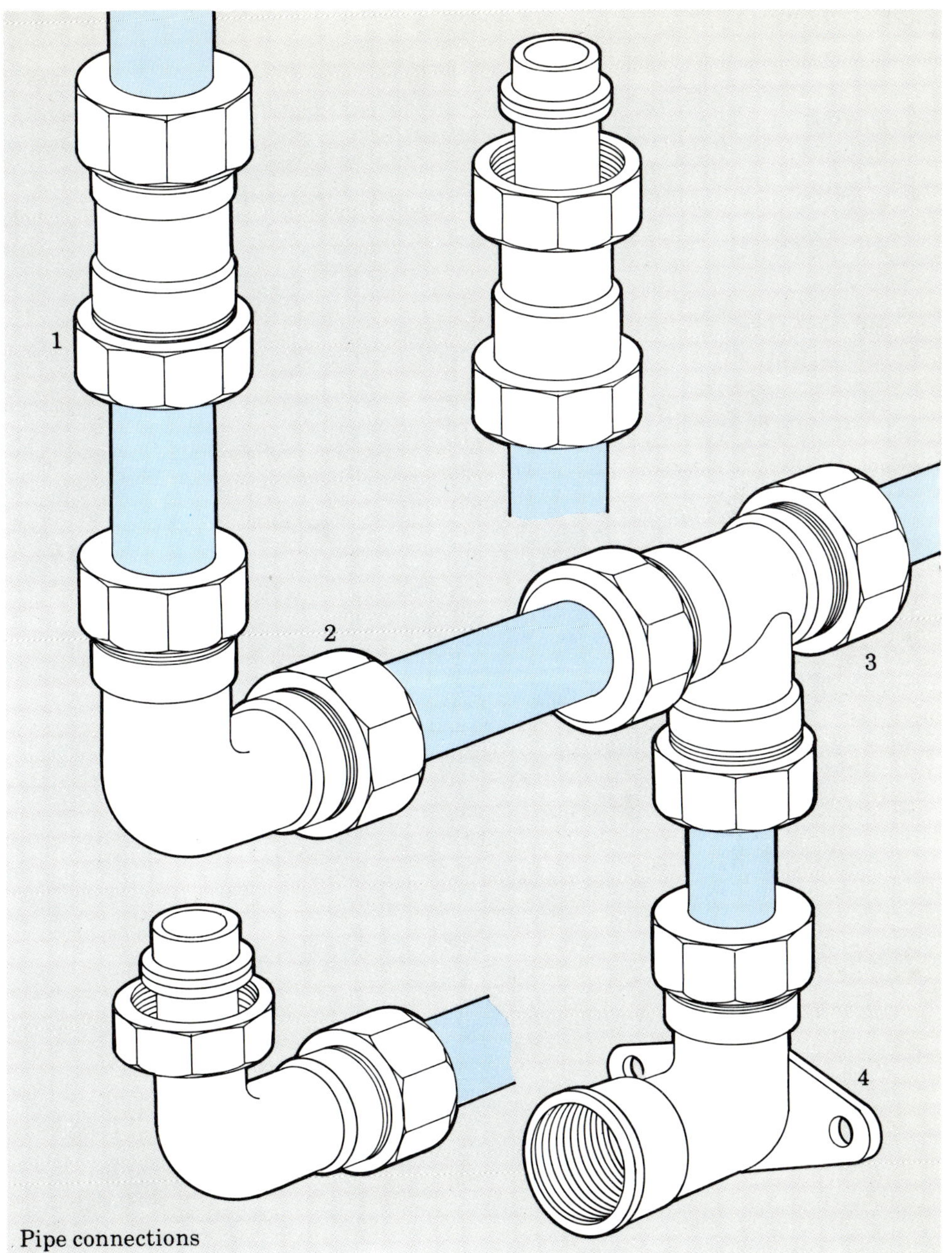

Pipe connections

Where to find out more

Magazines are a regular source of information on techniques, tools and materials: two valuable ones are *Do It Yourself* from newsagents which will answer questions from readers; and *Handyman Which* available to subscribers to *Which?* (the two combined cost £8 a year, from Consumers Association, 14 Buckingham Street, WC2), containing a lot of comparative tests on various brands of tools and materials.

If you have questions about the water supply, your own Water Authority should answer them. On problems concerning drains, go to the Environmental Health Officer of your local Council

To track down suppliers of materials or fittings you need, ask the National Federation of Builders' and Plumbers Merchants (15 Soho Square, W1V 5FB) for addresses of retailers in your own neighbourhood, or look in the Yellow Pages of the telephone directory.

A catalogue that gives you a good idea of the wide range of gear that can be hired is obtainable from Hire Service Shops, one of several chains in this field. Their head office is at 31 London Road, Reigate, Surrey.

Many big cities have Building Centres where you can see displays of taps, showers, basins and so forth, and obtain sheaves of brochures. Those in London, Cambridge, Bristol and Manchester have very good bookshops that sell do-it-yourself books.

Here is a list of addresses of some manufacturers whose products have been mentioned in this book. Most are willing to provide the names of local retailers or to supply you direct if there is no local retailer; and all can be asked for advice on problems within their field.

Dismantling lubricants 'Plus Gas' and 'WD40'
Fosmin Ltd., Baston Mills, Hyde, Cheshire;
Cadulac Ltd., Old Boston Estate, Haydock, Lancs.

Nut splitter
Barrus Ltd., 12 Brunel Road, W3 7UY.
Pipe wrenches, screwdrivers, etc.
Footprint Ltd., Hollis Croft, Sheffield.
Pliers
M. Mole Ltd., Newport, Gwent.
Washer/jumper unit ('Full Stop')
Celmac Ltd., Unit 3, Ferry Lane, Brentford, Mdx
TW8 0BG.
Supataps
Deltaflow Ltd., Manor Road, Crawley, Sussex.
Tap conversion kit
Aubit Ltd., Harrowbrook Estate, Hinckley, Leics.
LE10 3DF;
Longwick Ltd., Oakley House, Longwick, Bucks
(heads only).
Downpipe connector (for hoses)
Lockstoke Ltd., Easebourne Lane, Midhurst,
Sussex.
Downpipe connector (for water butt)
Intrend, 40 Church Road, Paddock Wood, Kent.
Epoxy Resin ('Isopon')
W. David Ltd., Northway House, High Street, N20 9LR.
Plastic Padding Ltd., Sands Estate, High
Wycombe, Bucks.
*Vinyl adhesive, polyurethane foam ('PVX',
'Fomofill')*
Sabey Simpson Ltd., 26 Portland Drive, Merstham,
Surrey.
Epoxy putty ('Sylmasta'), waterproof adhesive tape
Sylglas Ltd., Knights Hill, SE27.
Pipeseal ('Rotunda')
Copydex Ltd., 1 Torquay Street, W2.
Descalers
Albright & Wilson, Box 80, Oldbury B69 4LN;
Descalite Ltd., 60 Swaislands Drive, Crayford,
Kent.
Descaler (central heating – 'Fernex')
Industrial Services Ltd., 214 High Street,
Waltham Cross, Herts.

Shower kit
Deltaflow Ltd., 1 Hagley Road, Birmingham B16 8TG.
Sink water-heater
Heatrae Ltd., Hurricane Way, Norwich NR6 6EA.
Slim flush panel
Fordham Ltd., Melbourne Works, Dudley Road,
Wolverhampton.
Compression joints ('Conex')
Delta Ltd., Whitehall Road, Tipton DY4 7JU.
Bath panels and strip ('Easifit', 'Frostguard')
Chloride Shires Ltd., Guiseley LS20 8AP.
'Copperbend' pipe, compression joints for copper
pipes
Ucan Ltd., 11 Old Esher Road, Hersham, Surrey.
Robert Dyas Ltd., the chain of hardware shops,
stock many of the items mentioned.
Acid cleanser ('Disclean')
Southern Chemicals Ltd., 28 Courtenay Road,
Woking, Surrey.
Liquid solder ('Solderlene')
Lambert & Smyth, 19 Heathmans Road, SW6.
Rust preventatives and anti-rust paints
Jenolite Ltd., Rusham Road, Egham, Surrey;
A. Sanderson Ltd. ('Kurust'), Ropery Street, Hull
HU3 2BX;
BP Aquaseal Ltd., Kingsnorth, Hoo, Kent;
T. Ness Ltd. ('Presomet'), Nantgarw, Cardiff CF4 7YH.
Anti-condensation paint ('Korkon')
International Paints Ltd., 24 Canute Road,
Southampton SO9 3AS.
Epoxy bath enamel
Taylors Ltd., Victoria Works, Garside Street,
Bolton BL1 4AE;
Porcelain Newglaze Ltd., 1 Moreland Drive,
Gerrards Cross, Bucks;
Renubath Ltd., 596 Chiswick High Road, W4.
Washing-machine plumbing-in kit ('Siroflex')
Clothiers Ltd., Trading Estate, Addlestone,
Surrey.

Tap adaptors, roller base, hose clamp
Robimatic Ltd., Kingway House, North Parade,
Horsham, Sussex.
Gutter grid
Coburg Brush Ltd., Brook House, N. Brook Street,
Newbury, Berks.
Plastic steel, silicone rubber
Devcon Ltd., Sheale, Berks.
*Cement bandage, waterproof tape, silicone rubber,
joint sealant ('Rok-Rap', 'Flashband', 'Colourseal',
'Plumbers Mait')*
Evode Ltd., Stafford ST16 3EH.
Cement-acrylic filler
Moltofill Ltd., 30 Broadwater Road, Welwyn
Garden City, Herts.